Francis-99 Workshop 2: Transient Operation of Francis Turbines 2016

AA006817

Journal of Physics: Conference Series Volume 782

Lulea, Sweden
14 – 15 December 2016

Editors:

Michel J. Cervantes
Chirag Trivedi
O.G. Dahlhaug
T. Nielsen

ISBN: 978-1-5108-3533-7
ISSN: 1742-6588

Printed from e-media with permission by:

Curran Associates, Inc.
57 Morehouse Lane
Red Hook, NY 12571

Some format issues inherent in the e-media version may also appear in this print version.

Printed by Curran Associates, Inc. (2017)

For permission requests, please contact the Institute of Physics
at the address below.

Institute of Physics
Dirac House, Temple Back
Bristol BS1 6BE UK

Phone: 44 1 17 929 7481
Fax: 44 1 17 920 0979

techtracking@iop.org

Additional copies of this publication are available from:

Curran Associates, Inc.
57 Morehouse Lane
Red Hook, NY 12571 USA
Phone: 845-758-0400
Fax: 845-758-2633
Email: curran@proceedings.com
Web: www.proceedings.com

Table of contents

Volume 782

Francis-99 Workshop 2: transient operation of Francis turbines
14–15 December 2016, Lulea, Sweden

Accepted papers received: 2 December 2016Published online: 16 January 2017

Preface

Papers

IOP Conf. Series: Journal of Physics: Conf. Series **782** (2017) 011001 doi:10.1088/1742-6596/782/1/011001

PREFACE FRANCIS-99

Francis-99 is a set of workshops jointly organized by the Norwegian University of Science and Technology (NTNU), Norway and Luleå University of Technology (LTU), Sweden in the same spirit as the previous Turbine-99 workshops. The Francis-99 workshops aim to determine the state of the art of high head Francis turbine simulations (flow and structure) under steady and transient operating conditions as well as promote their development and knowledge dissemination openly. Three workshops are initially planned:

- Workshop 1: steady operation of Francis turbines (performed in December 2014)
- Workshop 2: transient operation of Francis turbines (December 2016)
- Workshop 3: FSI of Francis turbines (December 2018)

A high head Francis turbine model, named the Tokke model, has been designed and experimentally investigated at the Water Power Laboratory, NTNU. The complete geometry of the model and mesh are now freely available on the site http://www.ntnu.edu/nvks/francis-99 together with a large set of experimental pressure and velocity measurements. The organisers expect this geometry to become with time a reference test case to the hydraulic community for research and development on high head Francis turbines and the workshops a meeting place to discuss developments, potentials, issues... on a common and open test case.

The present proceeding contains the papers presented at the second workshop at NTNU the 14th and 15th of December 2015. A total of 10 papers were presented. A large variety of codes and models were used highlighting different issues in the simulation of high Francis turbines.

The editors:
Prof. Michel J. Cervantes (LTU, NTNU),
Dr. Chirag Trivedi (NTNU)
Prof O.G. Dahlhaug (NTNU)
Prof. T. Nielsen (NTNU)

IOP Conf. Series: Journal of Physics: Conf. Series **782** (2017) 011002 doi:10.1088/1742-6596/782/1/011002

Peer review statement

All papers published in this volume of *Journal of Physics: Conference Series* have been peer reviewed through processes administered by the proceedings Editors. Reviews were conducted by expert referees to the professional and scienti c standards expected of a proceedings journal published by IOP Publishing.

IOP Conf. Series: Journal of Physics: Conf. Series **782** (2017) 012001 doi:10.1088/1742-6596/782/1/012001

Francis-99: Transient CFD simulation of load changes and turbine shutdown in a model sized high-head Francis turbine

Peter Mössinger, Roland Jester-Zürker, Alexander Jung

Voith Hydro Holding GmbH & Co KG, St Pöltener Straße 43, 89522 Heidenheim, Germany

E-mail: peter.moessinger@voith.com

Abstract. With increasing requirements for hydropower plant operation due to intermittent renewable energy sources like wind and solar, numerical simulations of transient operations in hydraulic turbo machines become more important. As a continuation of the work performed for the first workshop which covered three steady operating conditions, in the present paper load changes and a shutdown procedure are investigated. The findings of previous studies are used to create a 360° model and compare measurements with simulation results for the operating points part load, high load and best efficiency. A mesh motion procedure is introduced, allowing to represent moving guide vanes for load changes from best efficiency to part load and high load. Additionally an automated re-mesh procedure is added for turbine shutdown to ensure reliable mesh quality during guide vane closing. All three transient operations are compared to PIV velocity measurements in the draft tube and pressure signals in the vaneless space. Simulation results of axial velocity distributions for all three steady operation points, during both load changes and for the shutdown correlated well with the measurement. An offset at vaneless space pressure is found to be a result of guide vane corrections for the simulation to ensure similar velocity fields. Short-time Fourier transformation indicating increasing amplitudes and frequencies at speed-no load conditions. Further studies will discuss the already measured start-up procedure and investigate the necessity to consider the hydraulic system dynamics upstream of the turbine by means of a 1D3D coupling between the 3D flow field and a 1D system model.

1. Introduction

An increasing demand for clean peak power energy leads to new requirements for hydraulic turbines such as a rising number of load changes or start-stop procedures. So far, numerical simulations are mostly based on the investigation of steady operating points. In recent years the focus shifted towards extreme operation ranges like deep part load and overload as well as load variations, start-stop conditions or total load rejection with the goal to better understand occurring local flow phenomena. This leads to new challenges for computational fluid dynamics (CFD) methods like the representation of moving guide vanes or to take account of hydrodynamics up- and downstream of the turbine. First studies were done by Nicolle et al. [1, 2], investigating a startup procedure or Li et al. [3] and Mössinger et al. [4] simulating a load rejection for a prototype Francis turbine. In case of a load rejection either the domain upstream the guide vanes has to be fully modeled or with the aid of a 1D Code considered in a simplified way. In the first of three upcoming workshops different numerical solver, mesh densities and simulation approaches were used to investigate three different operating points [5]. In the current, second workshop load changes and shutdown/startup procedures are investigated.

2. Experimental setup

2.1. Model machine

The test rig used for this measurement campaign is located at the Water Power Laboratory, Norwegian University of Science and Technology, Trondheim, Norway. The turbine, operating with an open loop circuit, is scaled down from prototype 1:5,1 [6]. Besides the runner with a total of 15 splitters and 15 full length blades the model includes 14 stay vanes inside the volute casing, 28 guide vanes and an elbow draft tube. Figure 1 shows the top view (left) and the meridional section (right) of the model machine. Line 1 and 2 are indicating the particle image velocimetry (PIV) measurement lines and DT5 the piezoelectric dynamic pressure sensor.

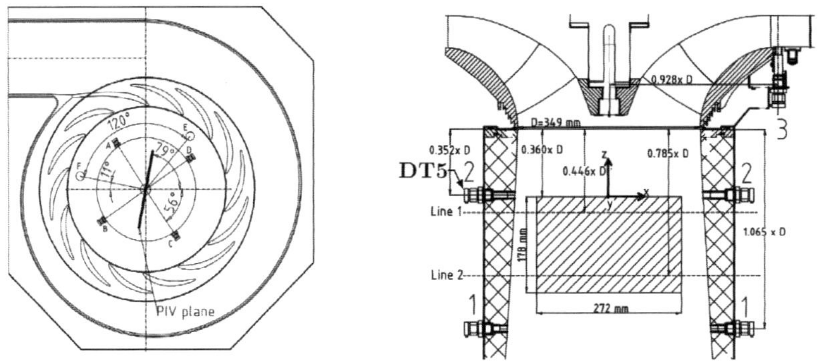

Figure 1: model machine, location of the pressure sensors and PIV measurement sections [7]

2.2. Measurement

Similar to the first workshop three different operating points are investigated. In contrast to previous investigations the part load (PL) and high load (HL) operation point are modified such that they have now a constant speed factor n_{ed} (cf. figure 2). Based on these conditions, load changes are initiated going from best efficiency point (BEP) to part load and high load. Axial and radial velocity measurements in the draft tube as well as pressure measurements in the spiral case, vaneless space and draft tube are performed. Table 1 summarizes the conditions for all three operating points.

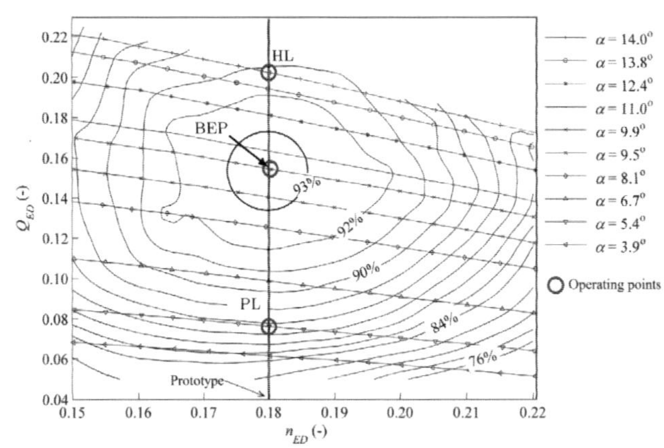

Figure 2: hill chart of the high head Francis turbine model [5]

Table 1: operating points and experimental data for pressure and velocity measurement

Parameter	symbol	unit	PL	BEP	HL
Net head	H	m	11,87	11,94	11,88
Flow rate	Q	m^3/s	0,14	0,20	0,24
Runner angular speed	n	1/min	332	332	332
Guide vane angle	α	°	6,72	9,84	12,43
Efficiency	η_h	%	90,13	92,39	91,71
Hydraulic torque	T	Nm	420,8	620,6	744,4
Speed factor	n_{ed}	-	0,179	0,178	0,179
Discharge factor	Q_{ed}	-	0,110	0,152	0,184

3. Numerical setting

3.1. Solution setup

All simulations are performed with the software ANSYS CFX v16.0. CFX is an established and well validated fluid solver and allows very high accuracy in the prediction of complex turbulent flows [8]. It uses a finite volume based discretization scheme up to second order accuracy for convective fluxes and truly second order accuracy for diffusive fluxes. Time step for all simulations is $\Delta t = 2 \cdot 10^{-3}$ s which is equivalent to a runner rotation of $\Delta_{RU} = 4°$ per time step. Convergence is achieved with root mean square residuals of pressure and mass-momentum of $r \leq 2 \cdot 10^{-5}$ with minimum seven iterations per time-step. Turbulent scales were modeled by the SST k-ϵ-model with automatic wall function.

3.2. Mesh and boundary conditions

Based on the outcome of the first workshop an appropriated mesh density is used [9]. All simulations were done with a 360° domain, including the volute casing (SC), all 14 stay vanes (SV), 28 guide vanes (GV), 30 runner blades (RU) and the elbow draft tube (DT). Table 2 summarizes the resulting mesh and quality parameters. Non matching grid interfaces are modeled with the general grid interface (GGI) where the transition from stationary to rotating domain and vice versa is handled with a transient rotor stator interface, accounting for all transient flow characteristics. Meshing is done with support of an in-house tool, using scaled geometry and block-structured hexahedral cells, ensuring constant mesh conditions when varying the guide vane angle and furthermore allowing a rapid change of mesh properties or interface positions. The stay vanes in the present geometry extend far in the spiral case which prevents a separate modeling as individual domains. Therefore the spiral case and the stay vanes are

Table 2: used grid densities and resulting quality parameters

	SC+SV	GV	RU	DT
Nodes ($\times 10^6$)	4,13	14,46	9,57	1,63
Total nodes ($\times 10^6$)		29,8		
Min. element angle	13,3°	15,0°	15,1°	57,3°
Max. aspect ratio	280	148	255	55
Max. expansion factor	39	84	49	3
Mean y^+ (BEP)	50	23	60	49

IOP Conf. Series: Journal of Physics: Conf. Series **782** (2017) 012001 doi:10.1088/1742-6596/782/1/012001

meshed manually as a single domain. Since the discharge and inlet pressure measurements show some time delay, a linear discharge variation based on the end points of the transitions is imposed at the inlet. The domain outlet is defined with a zero static pressure condition allowing backflow by switching to total pressure based on the normal component of velocity.

3.3. Guide vane rotation

Moving guide vanes during load changes and shutdown are realized with a combination of node motion and automated remeshing. ANSYS CFX already comes with the ability to deform meshes [10]. While the boundary movement is specified, all remaining nodes shift according to a displacement diffusion by solving the equation:

$$\nabla \left(\Gamma_{disp} \nabla \delta\right) = 0 \tag{1}$$

Mesh stiffness Γ_{disp} increases as the control volume size decreases:

$$\Gamma_{disp} = \left(\frac{\forall_{ref}}{\forall}\right)^{C_{stiff}} \tag{2}$$

For this case, larger control volumes absorb more mesh motion. An increase of the stiffness model exponent C_{stiff} from 2 (default) to 15 yields to a much more abrupt stiffness variation, ensuring a desirable mesh quality along the guide vane movement. As the guide vane movement is not a linear function, measurement values were taken for the simulation. As shown in figure 3 a moving mesh approach is sufficient for both load changes where the minimum mesh angles are kept in an acceptable range.

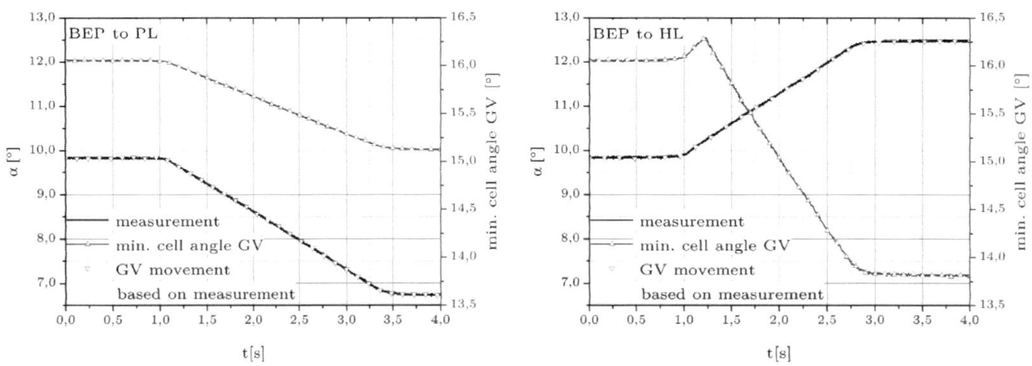

Figure 3: guide vane angle and minimum cell angle for load changes from best efficiency to part load (left) and high load (right)

However for a shutdown process the mesh quality successively decreases as the guide vane angle goes to zero and thus a mesh adaption is required. Therefore an automated remesh routine is used [4]. Depending on some logical expression such as maximum runtime, minimum cell angle, maximum aspect ratio or any combination, the simulation is paused and with the aid of an in-house mesher, depending on the current guide vane angle, a new mesh is generated. After loading the mesh into CFX and interpolating the old results, the simulation proceeds. This procedure is valid up to small openings ($\alpha < 2°$) where manually generated meshes are necessary. Figure 4 shows the guide vane movement, corresponding minimum cell angles and maximum aspect ratios. At a run-time of 5,6 s and 7,4 s new meshes were generated.

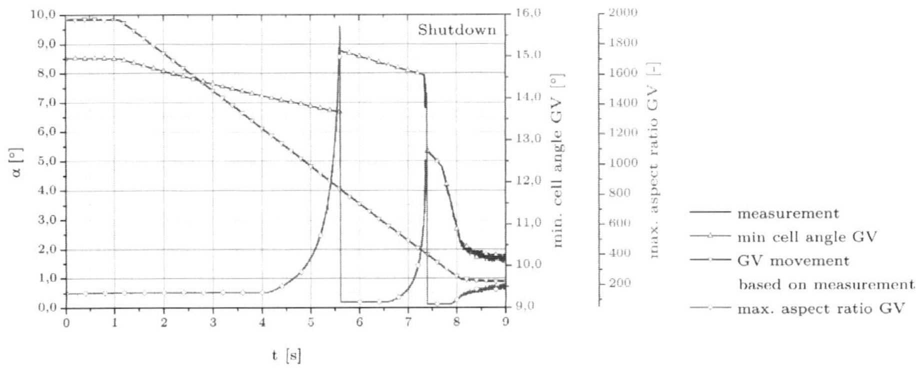

Figure 4: guide vane angle, min. cell angle and max. aspect ratio for shutdown procedure

4. Results

4.1. Operating points

In a first step three operating points that serve as start and end points of the load changes are investigated. Figure 5 shows the comparison between measurement, averaged over 5 s and simulation results which are averaged over 3 s. As in the first workshop the guide vane angle is corrected to ensure the similarity between the velocity fields up- and downstream of the runner for the experiment and the simulation. However in contrast to previous investigations the total guide vane angle correction set at best efficiency is kept constant for all operating points to ensure identical GV rotation. All investigated values are within an acceptable range of less than 4 % deviation, except the pressure measurement at the vaneless space. As indicated in the small sketch the pressure position is at the suction side near the trailing edge. A positive correction of the guide vanes will decrease the pressure and therefore lead to an underestimation compared to the measurement.

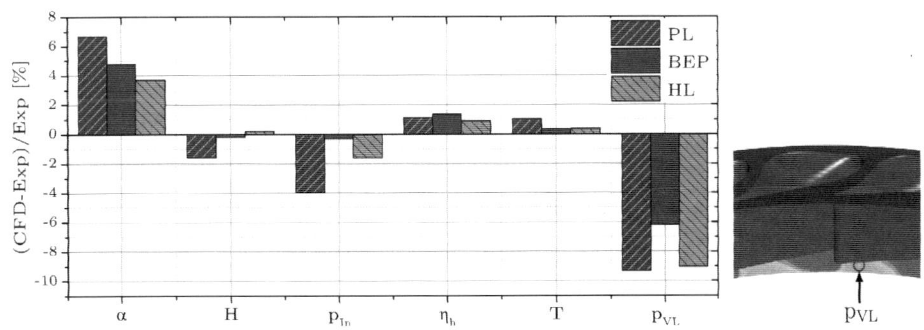

Figure 5: comparison between numerical and experimental values for all three operating points

Axial velocities of the simulation results at the top evaluation plane in the draft tube are in good agreement to the measurement for all three operating points (cf. figure 6), where positive values are pointing in flow direction. At best efficiency, velocities in the draft tube center are slightly underestimated where at part load the simulation shows some backflow which is also an underestimation compared to the measurement. The averaged velocity profile for part load condition after the load change from best efficiency was added to the left diagram of figure 6. Slight deviations between the measured steady operating point and the velocity profile after the load change are visible. This was only recognizable for part load conditions.

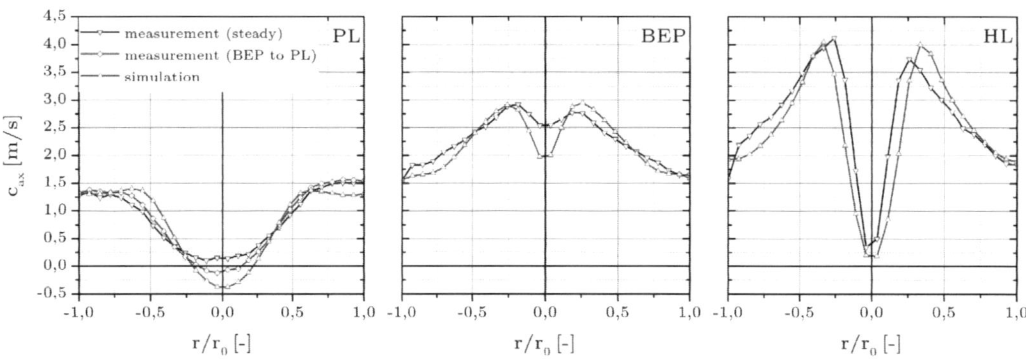

Figure 6: averaged axial velocity distribution at top evaluation plane for part load (left), best efficiency (center) and high load (right)

4.2. Load changes

4.2.1. Best efficiency to part load The first load change starts from best efficiency and moves towards part load. There are only minor differences between the top and bottom measurement plane, therefore a representation is focused on the top plane. In figure 8a a comparison of the axial velocity at four different time instances during the load change is shown, where in figure 8b four different positions in radial direction are displayed. The reduced flow velocity in the draft tube center is visible, where the simulation slightly underestimates the velocity in the main area. Also over time a decreasing axial velocity is visible and in good agreement to the measurement.

In figure 7 time and spacial domain are combined in one contour plot. Along the abscissa the axial velocity at one time instance is plotted, where the ordinate shows the velocity at one measurement point along time. Qualitative, a good agreement between measurement and simulation is visible. At the beginning higher velocities are present near the draft tube cone and during the load change to part load the velocity in the draft tube center decreases.

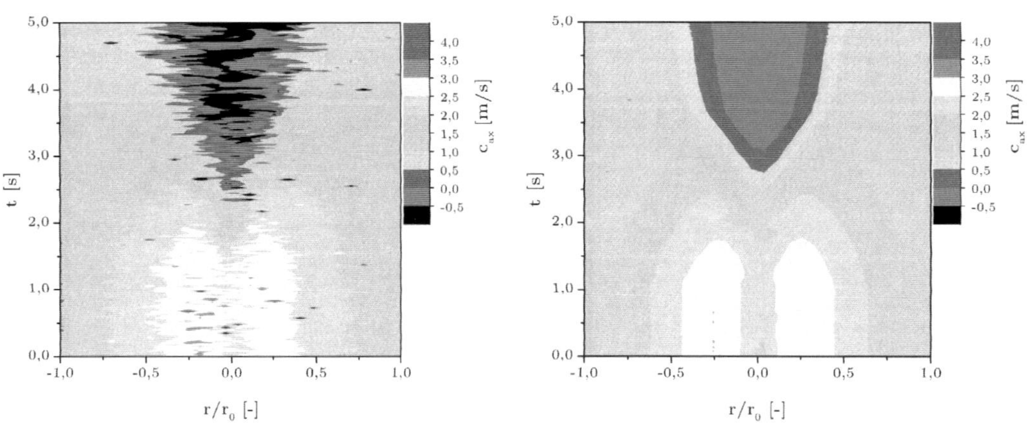

Figure 7: axial velocity distribution at top evaluation plane of measurement (left) and simulation (right) for load change from best efficiency to part load

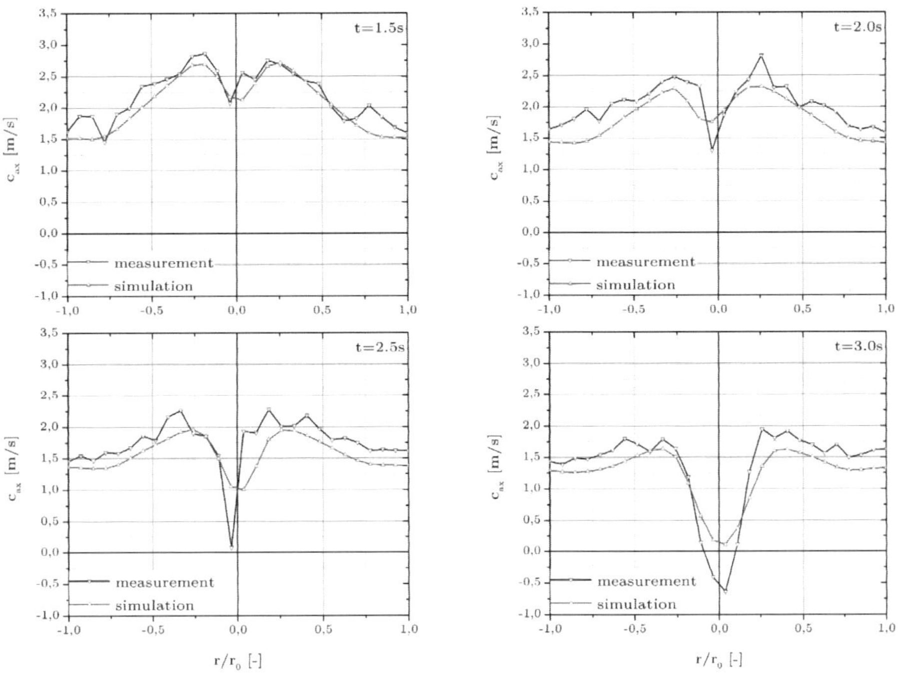

(a) axial velocity distribution at four different times

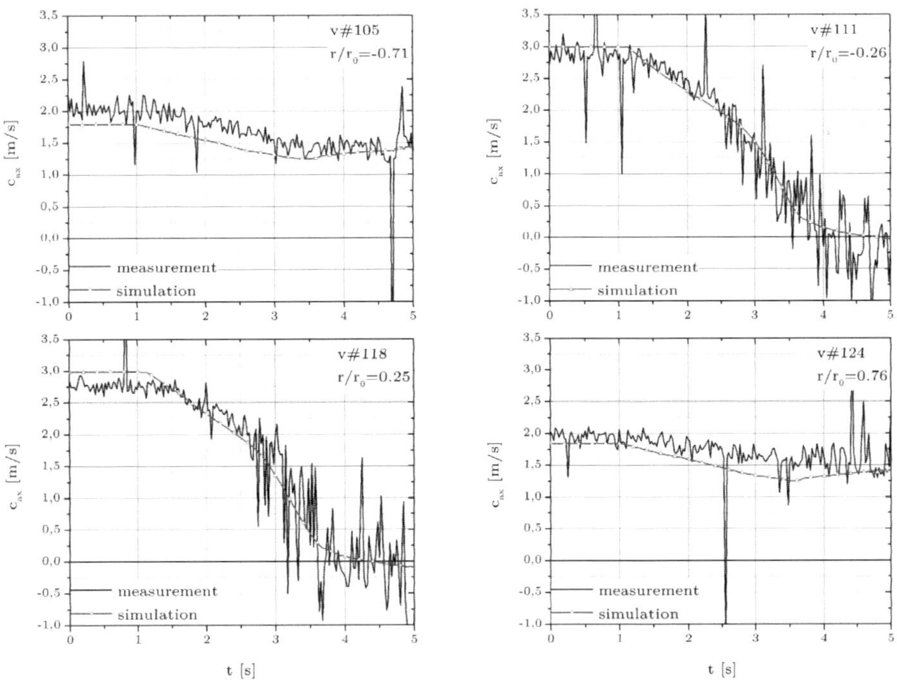

(b) axial velocity distribution at four different radial positions

Figure 8: axial velocity distribution at top evaluation plane for load change from best efficiency to part load at four different times and four different radial positions

IOP Conf. Series: Journal of Physics: Conf. Series **782** (2017) 012001 doi:10.1088/1742-6596/782/1/012001

4.2.2. Best efficiency to high load For the load change from best efficiency to high load the fluctuations of measured velocities are increasing, making a direct comparison of different points in the draft tube more difficult. But again a contour plot of the axial velocity can be plotted (cf. figure 9). Both contour plots showing again a good conformity. Towards high load the velocities sidewards of the draft tube cone are increasing, where in the center a stagnation region is visible.

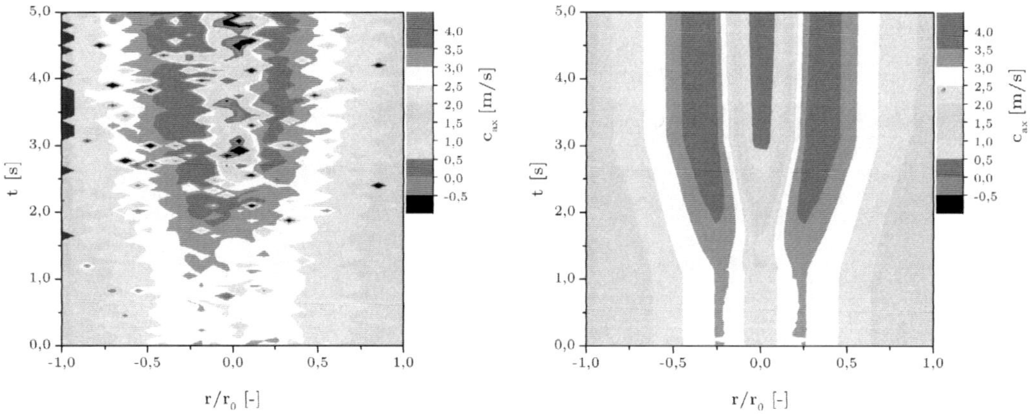

Figure 9: axial velocity distribution at top evaluation plane of measurement (left) and simulation (right) for load change from best efficiency to high load

4.3. Turbine shutdown

In a last step a shutdown procedure is initiated. While the runner speed is kept constant the guide vanes are closing to speed-no-load conditions at $\alpha = 0,9°$. Similar to previous investigations at least one second of simulation time before and after the guide vane movement is added to reach steady conditions. For the shutdown this results in a total simulation time of 9 s. Figure 10 shows again a comparison between measurement and simulation results for the axial velocity at four different time instances (figure 10a) and radial positions (figure 10b). Again the temporal and spacial agreement between both is satisfying. Increasing velocity fluctuations at small openings are visible and captured well by the simulation. Close to speed-no-load condition the axial velocity is in general slightly overestimated as seen for t=7,5 s in figure 10a and in the contour plot of figure 11.

As a last comparison the vaneless space pressure is investigated (cf. figure 12). Due to the adopted guide vane opening the pressure signal is normalized with best efficiency conditions. As expected the simulations show pressure surges at beginning and end of the guide vane movement/discharge variation. This cannot be seen in the measurement. Also the simulation is overestimating (load change to part load) or underestimating (load change to high load) the final pressure level. The assumption of a linear discharge variation may be a modeling error in the current simulations. An improvement could be made by defining an inlet total pressure according to the measurement and let the discharge a result of the head loss due to the current guide vane opening, or even consider the upstream hydrodynamics with a full 3D model or a simplified 1D3D coupling method.

Contrary to the first workshop the draft tube pressure sensors are of piezo electric-dynamic type and thus capture pressure fluctuations more accurately. To compare the pressure signals a procedure to obtain the instantaneous mean of the DT5 pressure fluctuations adapted from

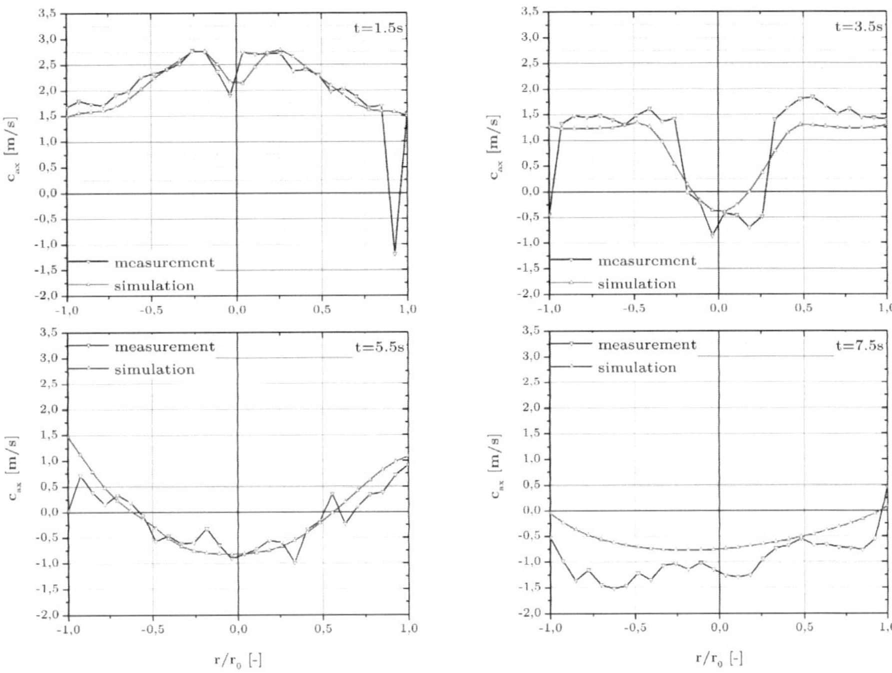

(a) axial velocity distribution at four different times

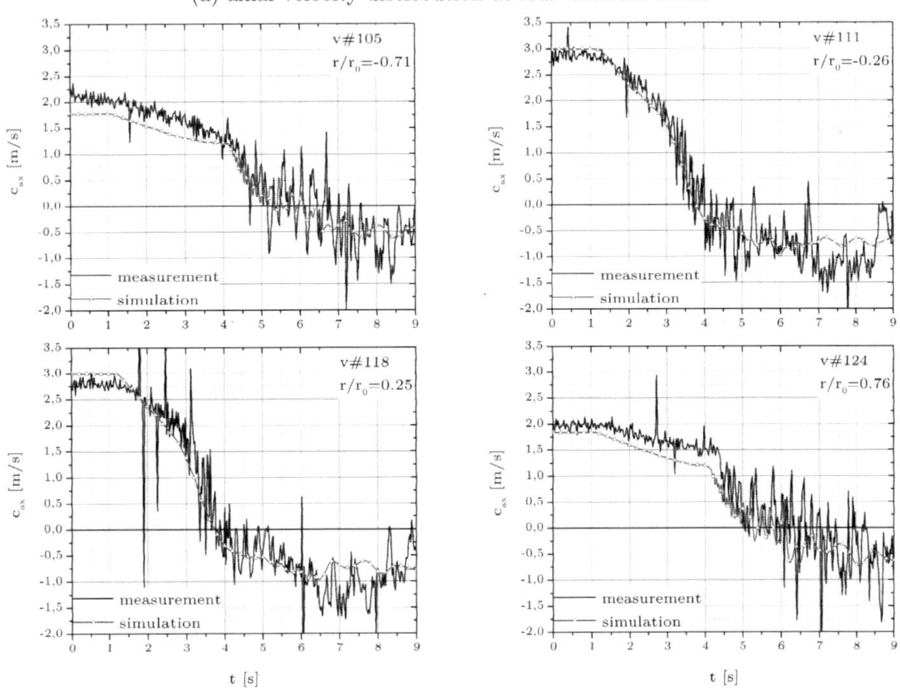

(b) axial velocity distribution at four different radial positions

Figure 10: axial velocity distribution at top evaluation plane for turbine shutdown at four different times and four different radial positions

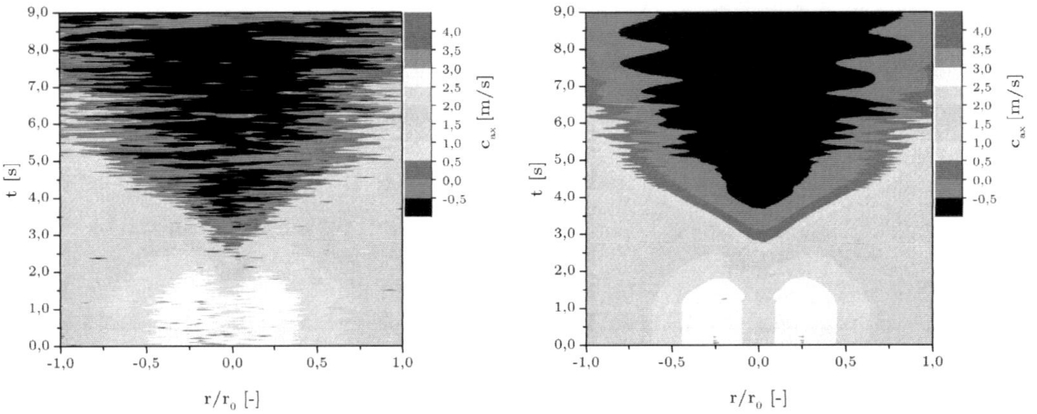

Figure 11: axial velocity distribution at top evaluation plane of measurement (left) and simulation (right) for turbine shutdown

related investigations are used [12]. Instead of the Savitzky-Golay filter a median filter is used to preserve instant changes and lower frequencies. To capture fluctuations of the pressure-time signal the instantaneous mean pressure ($\overline{p'}$) was then subtracted from the raw signal (p) and normalized with the energy at best efficiency condition.

$$\tilde{p} = p - \overline{p'} \tag{3}$$

$$\tilde{p_E} = \frac{\tilde{p}}{(\rho E)_{BEP}} \tag{4}$$

A limitation of the fast Fourier transformation (FFT) is to provide simultaneous time and frequency localization and therefore not very useful for analyzing time-variant, non stationary signals. To overcome this problem, the signal is segmented into narrow time intervals and take the Fourier transformation of each segment. For each FT segment simultaneous spectral informations of time and frequency are available, called short time Fourier transformation (STFT). However this method comes with the drawback of the so called uncertainty principle. Time resolution and frequency resolution can not be made arbitrary small at the same time

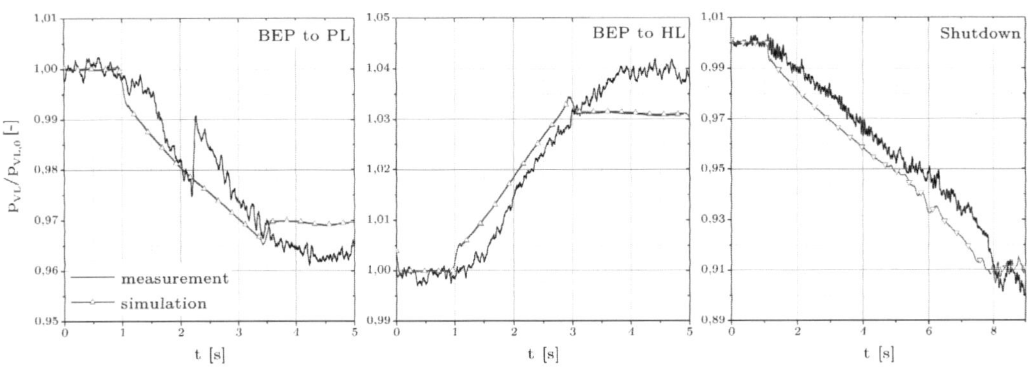

Figure 12: Normalized pressure at vaneless space for load changes from best efficiency to part load (left), to high load (center) and turbine shutdown (right)

and a compromise between both has to be chosen [13]. Figure 13 shows the amplitude pressure signal of the sensor DT5 in the draft tube extracted according to eq (3) and (4) as well as the resulting short time Fourier transformation for measurement and simulation.

For the measurement at 166 Hz ($30 \cdot f_n$) the rotor frequency is visible, where higher amplitudes at 40 Hz may be related to system exciting frequency [14]. At small openings perturbations at the draft tube measurement positions are increasing in intensity and frequency. Similar observations can be made for the simulation. As already seen in the vaneless space pressure signal, start and stop of the discharge variation are also introducing a pressure surge with oscillations, clearly visible at 1 s and 8,1 s. Increasing pressure amplitudes at $5 - 7$ s and to a certain range at speed-no-load condition (>8 s) are captured by the simulation. The absence of higher frequencies after 5,6 s seems to be mesh related since the first re-mesh process was made at this position. Small mesh adaption was necessary to maintain a good mesh quality. This has to be investigated in further studies.

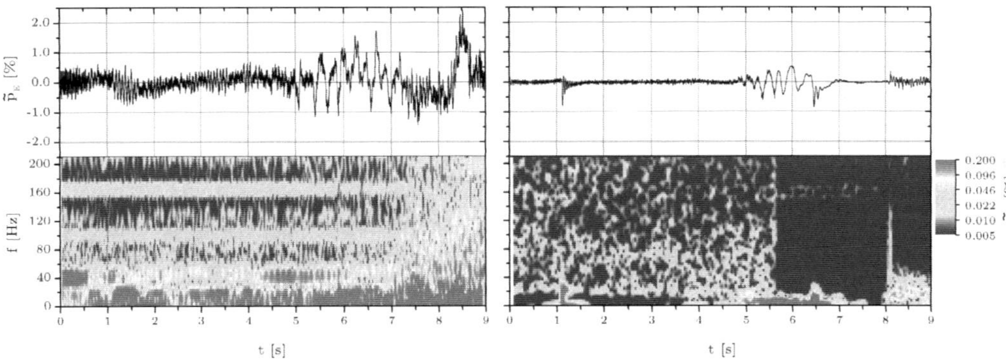

Figure 13: pressure amplitude and STFT at draft tube pressure location DT5 of turbine shutdown for measurement (left) and simulation (right)

5. Discussion

In the second stage of the Francis-99 workshop campaign numerical simulations of two load changes and a turbine shutdown were investigated. Based on the first workshop the most promising configuration was used to set up a 360° mesh including volute casing and draft tube. A moving mesh procedure with an automated remesh process was introduced to represent moving guide vanes up to small openings. Due to some low time response of discharge and pressure inlet measurement a linear discharge variation together with moving guide vanes were set as boundary conditions. First three steady operating points, best efficiency, part load and high load, were investigated and showed good agreement compared to the measurement. Change of axial velocity profiles during load changes starting from best efficiency going to part load and high load are captured well by the simulation. Also for the shutdown process the agreement was sufficient, where for small openings the velocities in general are slightly overestimated. Pressure signals of the simulation in the vaneless space and the draft tube are showing pressure surges at beginning and end of the discharge variation. This could not be seen in the measurement. Further studies have to improve the boundary conditions, to let the discharge variation be a result of the simulation. This can be achieved with a head variation according to the measurement or a constant upper reservoir pressure were the geometry upstream the turbine have to be fully modeled or considered with a simplified 1D model and coupled to the 3D domain. Also mesh dependency was seen for the frequency domain which has to be further investigated. As a last step starting from the end point of the shutdown process an already measured startup procedure will be modeled.

Nomenclature

Term	*Symbol*	*Unit*	**Indices**	
Stiffness model exponent	C_{stiff}	-	0	Normalized
Velocity, axial	c_{ax}	m/s	E	Specific energy
Runner angular speed	n	1/min	**Abbreviations**	
Speed factor	n_{ed}	-	BEP	Best efficiency
Pressure	p	Pa	DT	Draft tube
Inst. mean pressure	$\overline{p'}$	Pa	FT	Fourier transformation
Pressure amplitude	$\tilde{p}$	Pa	FFT	Fast Fourier transformation
Discharge	Q	m^3/s	GV	Guide vanes
Discharge factor	Q_{ed}	-	HL	High load
Dimensionless wall distance	y^+	-	PL	Part load
Guide vane angle	α	°	RU	Runner
Displacement	δ	m	SC	Spiral case
Hydraulic efficiency	η_h	%	SST	Shear stress turbulence model
Mesh stiffness	Γ_{disp}	-	STFT	Short time FT
Control volume size	$\forall$	m^3	SV	Stay vanes

References

[1] Nicolle J, Morissette J and Giroux, A 2012 Transient CFD simulation of a Francis turbine startup *IOP Conference Series: Earth and Environmental Science*

[2] Nicolle J, Giroux A and Morissette, J 2014 CFD configuration for hydraulic turbine startup *Proc. of the Hydraulic Machinery and Systems 27th IAHR Symposium* Montreal

[3] Li J, Yu J and Wu Y 2010 3D unsteady turbulent simulations of transients of the Francis turbine *Proc. of the Hydraulic Machinery and Systems 25th IAHR Symposium* Montreal

[4] Mössinger P and Jung A 2016 Transient two-phase CFD simulation of overload operating conditions and load rejection in a prototype sized Francis turbine *Proc. of the Hydraulic Machinery and Systems 28th IAHR Symposium* Grenoble

[5] Trivedi C, Cervantes M and Dahlhaug O 2016 Numerical techniques applied to hydraulic turbines: a perspective review *Applied Mechanics Review* **68**

[6] Trivedi C, Cervantes M, Ghandi B and Dahlhaug O 2013 Experimental and numerical studies for a high head Francis turbine at several operating points *J. of Fluid Engineering* **135**

[7] Cervantes M Second Francis-99 2016 Workshop *Norwegian University of Science and Technology, http://www.ntnu.edu/nvks/second-workshop* accessed August 22, 2016

[8] Aschenbrenner T, Otto A and Moser W 2006 Classification of vortex and cavitation phenomena and assessment of CFD prediction capabilities *Proc. of the Hydraulic Machinery and Systems 23rd IAHR Symposium* Yokohama

[9] Mössinger P, Jester-Zürker R and Jung A 2015 Investigation of different simulation approaches on a high-head Francis turbine and comparison with model test data: Francis-99 *J. of Physics: Conference Series* **579**

[10] ANSYS Inc: *ANSYS CFX-Solver Theory Guide, release 16.0*, 2015

[11] Trivedi C, Cervantes M, Ghandi B and Dahlhaug O 2014 Transient pressure measurements on a high head model Francis turbine during emergency shutdown, total load rejection and runaway *J. of Fluid Engineering* **136**

[12] Trivedi C, Cervantes M, Dahlhaug O and Ghandi B 2015 Experimental investigation of a high head Francis turbine during spin-no-load conditions *J. of Fluid Engineering* **137**

[13] Kıymık K, Güler İ, Dizibüyük A, and Akın M 2005 Comparison of STFT and wavelet transform methods in determinig epileptic seizure activity in EEG signals for real-time application *Computers in Biology and Medicine* **35**

[14] Trivedi C, Cervantes J.M, Ghandi B.K and Dahlhaug O.G 2014 Pressure measurement on high-head Francis turbine during load acceptance and rejection *J. of Hydraulic Research* **52** 283-97

IOP Conf. Series: Journal of Physics: Conf. Series **782** (2017) 012002 doi:10.1088/1742-6596/782/1/012002

CFD simulations of transient load change on a high head Francis turbine

Ken-Robert G. Jakobsen[1], Martin Aasved Holst[2]

[1] Senior Technical Consultant, EDRMedeso AS
[2] Manager – Technical Services, EDRMedeso AS

EDRMedeso AS, Leif Tronstads plass 4, 1337 Sandvika, NORWAY

ken-robert.g.jakobsen@edrmedeso.com, martin.aasved.holst@edrmedeso.com

Abstract. Motivated by the importance of better understanding the structural integrity of high-head hydraulic turbines operating at intermittent conditions, complete 360° steady-state and transient simulations of a Francis turbine are presented in this paper. The main target of the work has been to investigate different numerical approaches such as mesh deformation for different operating conditions. Steady-state simulations were performed at the best efficiency point (BEP) and used as initial conditions for the transient simulations considering load rejection from BEP to part load (BEP2PL) and during load acceptance from BEP to high load (BEP2HL). Simulation results were compared with experimental data available for the Francis-99 project where close agreement was found for the mesh independent solution. The transient load analyses showed general trends in accordance with the measurement reports, especially for the pressure in vaneless space that is of high importance regarding RSI effects. Some deviations were identified for the net head at load rejection for which further investigations will be conducted. All CFD simulations were performed at model scale with ANSYS CFX v. 17 at either 96 or 120 cores (2.60 GHz). The immersed boundary technique was tested during the initial stages of the project, but had to be abandoned due to severe memory requirements. Pressure amplitudes and other instantaneous results were not considered.

Abbreviations

- BEP - Best Efficiency Point
- BEP2HL - Load acceptance
- BEP2PL - Load rejection
- CFD - Computational Fluid Dynamics
- GV - Guide Vane
- HL - High Load
- NTNU - Norwegian University of Science and Technology
- PL - Part Load
- RANS - Reynolds Averaged Navier-Stokes Equations
- RSI - Rotor-Stator Interaction
- RPM - Revolutions Per Minute
- SST - Shear Stress Transport

1. Introduction

The work presented here has been conducted as part of the HiFrancis project and specifically aiming for the second Francis-99 workshop where the focus lays on steady and transient numerical simulations of the Francis-99 hydraulic turbine located at the Water Power Laboratory at NTNU. Measurements of several experimental studies have been made available for comparison with the numerical results obtained by the workshop participants [1]. The Francis-99 project consists of a total of three strongly coupled workshops, the final one scheduled to take place in 2018. An in-depth description of the problem at hand together with an overview of the existing experimental and numerical findings from the Francis-99 workshops conducted so far can be found in [2].

The future hydropower systems will require operations with increased flexibility due to the introduction of intermittent energy sources such as wind and solar. This includes large and fast ramping rates and more frequent load changes that must be part of the future design criteria for new high head Francis turbines as the systems will be subjected to higher stress levels and consequently more fatigue. More efficient and more accurate numerical simulations will help providing better insight into how these new load histories affect the structural integration of the turbines. A necessary aid in this work is the ever-increasing computer power providing both faster results and the ability to run more expensive simulations. The latter has also been one of the main motives for the present study considering transient CFD simulations of the complete 360° Francis-99 turbine. These types of analyses are rare due to their high level of computational intensity and will in that sense hopefully bring new and fruitful information to the work aiming at better understanding the Francis turbine.

A thorough overview of the status of numerical modeling and simulation of hydraulic turbines at transient operating conditions today is given in [3].

2. Model setup

All simulations presented in this document were performed on a three-dimensional computational domain with a 360° turbine where the fluid behavior was obtained by solving the transient Reynolds averaged Navier-Stokes equations. The software used was ANSYS CFX v. 17.

Figure 1 shows the computational domain obtained by extracting the confined fluid domain from the Francis-99 turbine geometry. The main boundary conditions driving the fluid flow have been indicated:

- Inlet: Mass flow.
- Outlet: Pressure outlet.
- Walls: No-slip.
- Interface type: Frozen and transient rotor.

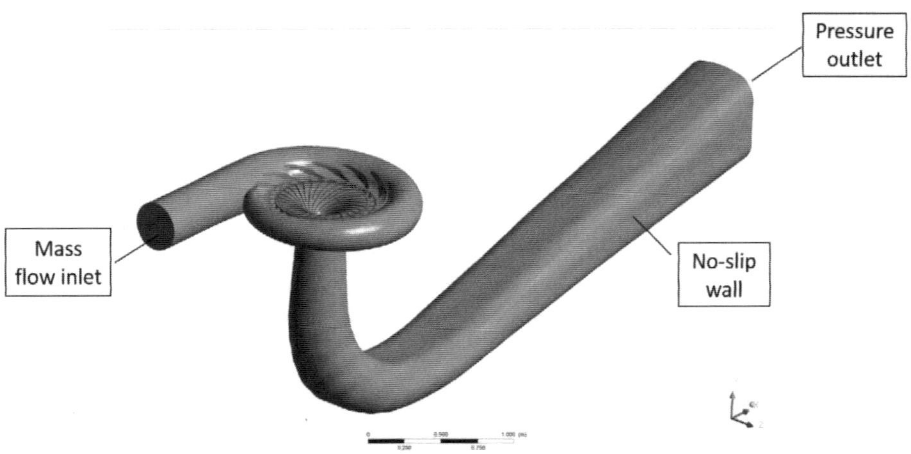

Figure 1. Computational domain.

In addition, the leakage gaps leading to the labyrinth seals were included by applying mass flow outlets at those locations. The latter was estimated to be 0.2% of the mass flow at BEP [4]. In [5], further details on the effect of including the leakage gaps can be found.

The working fluid in the analyses was incompressible water with density equal 999.8 [kg/m^3] and kinematic viscosity of 9.57e-7 [m^2/s].

2.1. Computational mesh

All simulations were performed on spiral casing meshes consisting of tetrahedral elements in the bulk flow and hexahedral elements in the near-wall regions capturing the boundary layer effects. On a general note, more tetrahedral elements are required compared with hexahedral elements to achieve mesh independency. The upside, however, is that they are very flexible and will easily mesh complex structures such as the present configuration. The latter is also a great advantage for the transient load scenarios where the rotation of the guide vanes were obtained utilizing mesh deformation following the specifics given in Table 1. It should also be noted that the immersed boundary method was tested for which the geometries subjected to motion are represented by momentum sources rather than resolved by computationally expensive meshes. However, this approach was abandoned due to very high memory requirements experienced with the present software.

As this work also aims at providing simulation procedures that are easily reproducible and at a reasonable cost post simulation, it was decided to focus solely on spiral casing meshes made of tetrahedrons, visualized in Figure 2. In the runner and draft tube, hexahedral elements were utilized for all cases. Details on mesh resolution are given in chapter 3.

Table 1. Mesh motion model.

Model	Option
Mesh motion	Displacement diffusion
Mesh stiffness	Increase near small volumes (model exponent = 1)
Guide vane displacement	Specified displacement (cylindrical components)

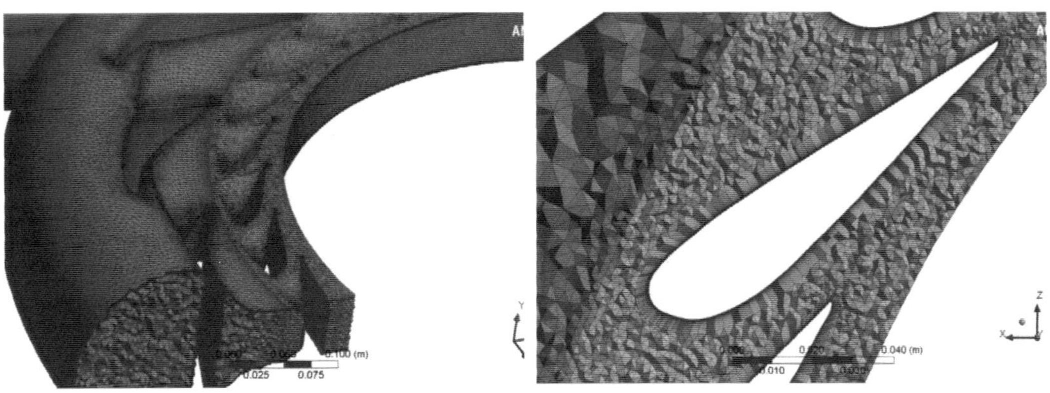

Figure 2. Computational domain with mesh consisting of tetrahedral elements (Case 3/8, see chapter 3). Left: Mesh at part of volute. Right: Boundary layer at one guide vane shown at a horizontal section.

2.2. *Turbulence model*

The industry standard SST turbulence model was applied to all simulations providing an accurate near-wall flow solution through the traditional k-omega model as well as in the free stream using the k-epsilon model with additional transportation term for shear stress. For this mixing to take place, the SST model does indeed require a fine mesh resolution near the no-slip boundaries and chapter 4. shows that this was accounted for in the present simulations by adjusting the first node from the wall (y+).

3. Simulation cases

The simulation cases performed in the present study have been listed in Table 2, covering both the mesh independence study and comparison with experiments (Case 1 – 9) as well as transient load cases (Case 9 and 10). All simulations include the full three-dimensional solution of the Navier-Stokes equations with a 360° turbine in steady-state (frozen rotor) or transient (rotating rotor) configurations. In accordance with the procedures for the available experimental data [1], the transient simulations see a turbine operating at a constant rotational speed of 333rpm throughout the investigation, while the inlet mass flow for the transient load cases between stationary operation points follow the same history as defined below for the guide vane movement. The simulation time steps have been defined as function of number of guide vanes (28) and angular speed of turbine (ω):

$$dt = \frac{2\pi}{28\omega N} \quad (1)$$

where N was chosen to ensure a fixed number of turbine rotations. The transient load cases include guide vane rotation to model either *load acceptance* or *load reduction* defined as:

1. Load acceptance: Increase turbine output from BEP to HL by opening the guide vanes from 9.84 degrees (BEP2HL). Mass flow linearly increased from 0.2 m^3/s (BEP) to 0.24 m^3/s (HL).
2. Load reduction: Decrease turbine output power from BEP to PL by closing the guide vanes from 9.84 degrees (BEP2PL). Mass flow linearly decreased from 0.2 m^3/s (BEP) to 0.14 m^3/s (PL).

Table 2. Simulation cases.

Case #	Type	Number of nodes ($x10^6$)	Time step [s]
1	Steady-state	7	-
2	Steady-state	22	-
3	Steady-state	42	-
4	Steady-state	100	-
5	Transient	7	$1x10^{-4}$
6	Transient	22	$1x10^{-4}$
7	Transient	42	$1x10^{-4}$
8	Transient	100	$1x10^{-4}$
9	Transient, BEP2HL	42	$8x10^{-4}$
10	Transient, BEP2PL	42	$8x10^{-4}$

The angular velocity of the guide vanes was derived from the operation scheme used in the experiments (see Figure 3) supplemented by their notes in [1] stating that: "At time t =1 s, the guide vanes were set to open/close to perform transient conditions. After certain time, the guide vanes reached to set value of angle (i.e., 6.72° or 12.43°) then steady state condition was followed".

Resulting from this, the load acceptance and rejection had to be run for operational times of 1.8s and 2.5s, respectively. In comparison, the BEP transient simulation was run for 0.54s, or three runner rotations, after which steady pressure signals were achieved after approximately half a rotation. The BEP transient result file was used as starting point for the load acceptance/rejection cases.

Briefly anticipating the course of the events, the final transient load cases defined in Table 2 were performed on the computational mesh consisting of 42 million cells as this gave mesh independent results. Further details are given in the subsequent chapters.

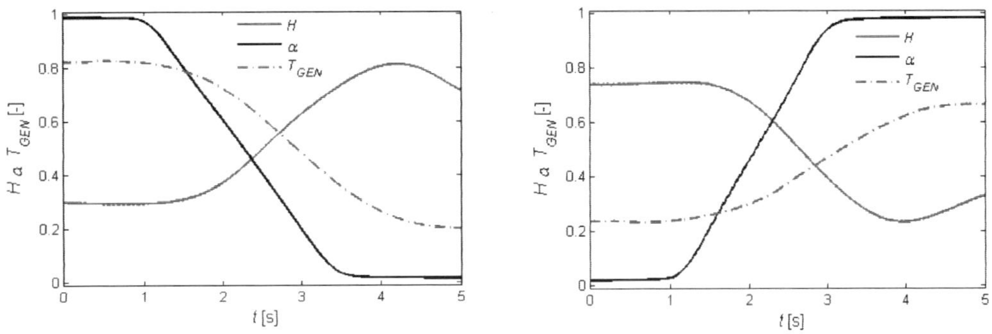

Figure 3. Operating schemes of guide vanes (α) in experiments [1]. Left: Load reduction (BEP2PL). Right: Load acceptance (BEP2HL).

4. Analysis results

The present chapter includes the analysis results obtained from the simulation cases defined in Table 2. These are presented in the following fashion:

1. Mesh sensitivity study
2. Comparison with experiments
3. Transient load change.

In accordance to the available experimental data, the main results of interest have been the net head and generator torque. The net head includes the casing inlet pressure and draft tube pressure and is defined as [1]:

$$Net\ head = \frac{P_{1abs}-P_{2abs}}{\rho g} + \frac{v_1^2 - v_2^2}{2g} + z \tag{2}$$

4.1. Mesh sensitivity study

Ranging from approximately 7 to 100 million computational nodes, four different mesh sizes were tested for the BEP stationary operating condition. In general, the mesh study was performed by refining both the bulk flow and boundary layer (lower y+) for each case, except between Case 3 and 4 where only the bulk flow resolution was increased. This is summarized in Table 3. It is here the average y+ values in the respective computational domains that have been reported. Obviously, as y+ is velocity dependent, this value will vary greatly in such a complex flow path as seen in the present configuration. Using the global average value is thus considered a representative measure for the turbulence model of choice.

Table 3. Mesh resolutions.

Case #	Number of nodes ($*10^6$)	Average y+ Volute [-]	Average y+ Runner [-]	Average y+ Draft tube [-]
1	7.4	41.1	158.6	71.2
2	22.4	24.5	35.5	21.9
3	42.5	2.8	2.8	1.6
4	100.1	3.1	3.0	1.4

As seen in Table 4 and the corresponding convergence plot in Figure 4, the results for net head and torque indicate that mesh independency for the 360° turbine was obtained at 42 million nodes for the frozen rotor (Case 3). The same conclusion was reached in the transient rotor simulations in Case 5 to 8. Also included is a comparison between torque based on the torque definition and on change in angular momentum from the inlet to the outlet of the runner. The latter derives from the Euler turbine equations and is proportional with $U \cdot C_u$, where U is the velocity of the runner and C_u is the tangential velocity at the inlet/outlet of the runner. These parameters should converge on the same value and is used to measure whether the mesh captures the velocity field accurately or not.

Based on the above, it was decided to utilize the Case 3/7 mesh for the transient load simulations. Note, however, that the boundary layer was not refined for the final mesh resolution.

Table 4. Mesh sensitivity study (BEP)

Case #	Net head [Nm]	Torque to generator [Nm]
1	12.44	643.6
2	12.32	635.0
3	11.88	617.5
4	11.86	616.5
5*	12.58	652.0
6*	12.45	643.1
7*	11.99	624.2
8*	11.97	622.9

**Mass flow average*

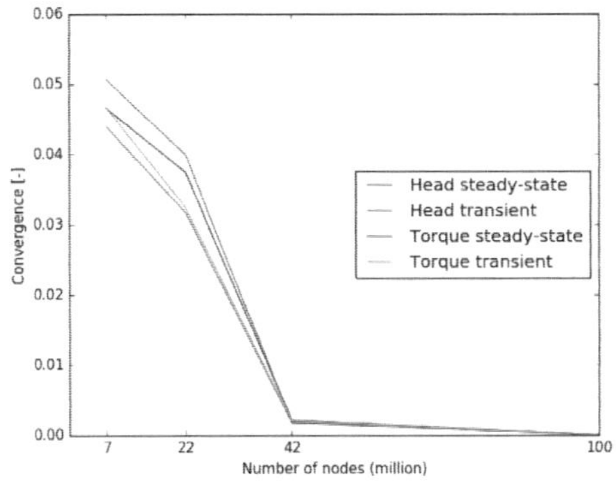

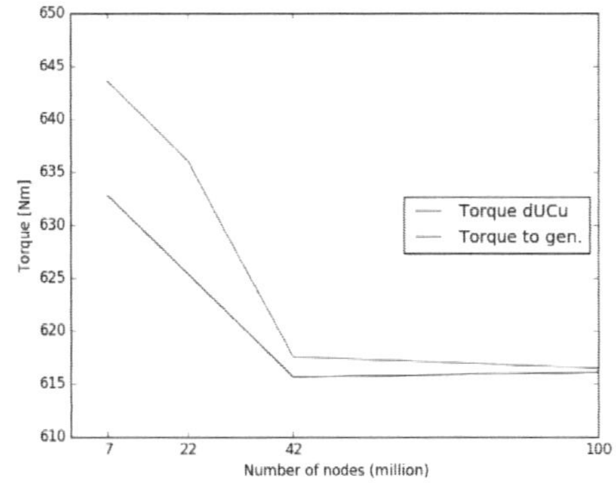

Figure 4. Left: Convergence plot of mesh sensitivity study for steady-state and transient simulation cases. Convergence is defined as $\left|\frac{\varphi_n}{\varphi_{100mill}} - 1\right|$, where φ is the variable in question. Right: Torque to gen. vs Torque ΔUC_u.

4.2. Comparison with experiments

Considering steady-state and transient simulations at the 42 million nodes mesh configuration (Case 3/7), Table 5 and Table 6 show the results obtained for both frozen and transient rotor together with the corresponding experimental values as extracted from [1]. The locations of the vaneless space probes (VL and VL2) are shown in Figure 5. In Figure 6, velocity and pressure contours for Case 7 have been presented.

In general, the analysis results show good agreement with the experimental data at all validation points for the steady-state simulations as well as when applying a transient rotor. Directly comparing the two methods, some variables improved slightly while other showed increased differences with the experiments when going from stationary to transient rotor. However, with deviations in the order of 0-1% except for the VL1 pressure in vaneless space at 3.8%, the overall accuracy was considered high and providing the necessary confidence in the analysis methodology of choice to continue with load acceptance and rejection as presented in the subsequent chapter.

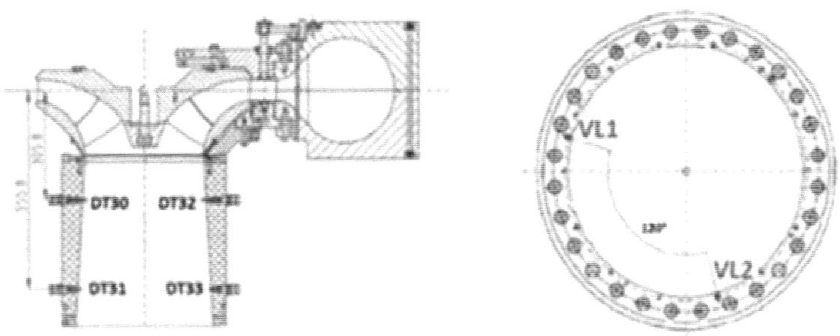

Figure 5. Location of pressure sensors. Only vaneless space (VL1 and VL2) considered here [6].

Table 5. Global measurement variables at BEP (steady-state and transient simulations).

Case #	Net head [m]		Torque to generator [Nm]		Inlet pressure [kPa]		Outlet pressure [kPa]		Hydraulic efficiency [%]	
Exp.	11.94	Dev. [%]	616.1	Dev. [%]	215.57	Dev. [%]	111.13	Dev. [%]	92.39	Dev. [%]
3	11.88	-0.5	617.5	0.2	217.95	1.1	111.83	0.6	92.41	0.0
7*	11.99	0.4	624.2	1.3	219.09	1.6	111.83	0.6	92.51	0.1

**Mass flow average values*

Table 6. Pressure in vaneless space at BEP (transient simulation).

Case #	VL1 [kPa]	VL2 [kPa]
Exp.	168.0	164.2
7*	174.3	164.2
Deviation [%]	3.8	0.0

**Mass flow average values*

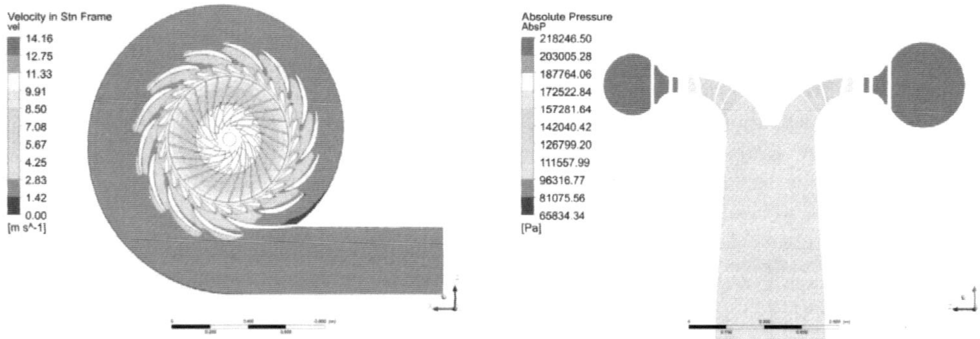

Figure 6. Case 7 results. Left: Velocity contours in stationary frame on horizontal mid plane. Right: Absolute pressure on vertical mid plane.

4.3. Transient load change

This chapter considers the load acceptance and load rejection analyses denoted as Case 9 (BEP2HL) and Case 10 (BEP2PL), respectively. As for the stationary load simulations, these results have also been subject to comparison with the experimental measurements in [1]. The transient rotor results obtained in Case 7 were used as initial values and the simulations were terminated at the time where the guide vanes found their final position. Note also that the mass flow specified at the inlet saw a linear change with time during the simulations, following the operating scheme seen in Figure 3. A close-up of the mesh around one of the guide vanes is given in Figure 7 showing its final two positions after rotating approximately 3 degrees in either direction, i.e. post load acceptance and load rejection.

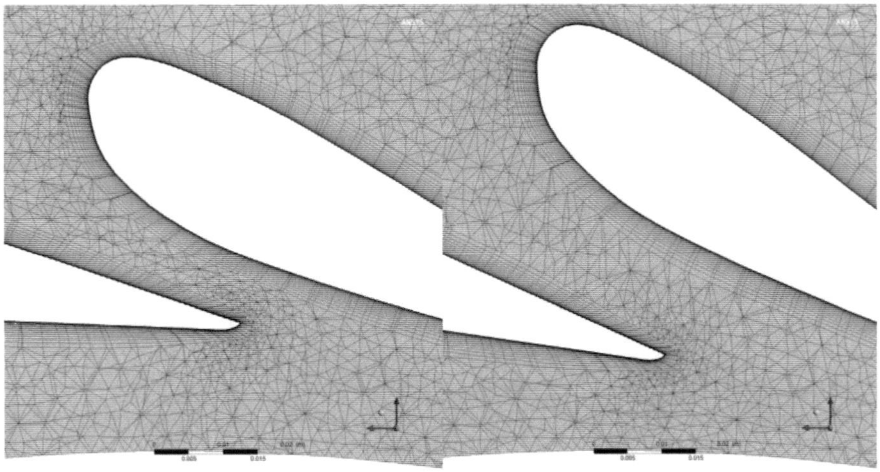

Figure 7. Final GV position during load rejection (left) and load acceptance (right).

Figure 8 and Figure 9 show the net head and the pressure in vaneless space (VL2) for load acceptance and load rejection, respectively. As amplitudes and instantaneous results have not been subject in the present section, second order polynomials were used to compare with the existing measurements as they provided the overall trends in the highly transient data pool. In agreement with the comparison data, the net head decreased during opening of the guide vanes (BEP2HL) and increased during load rejection (BEP2PL), while the VL2 pressures saw the opposite behavior. Clearly, the trends are matching with the experiments, especially for the VL2 pressure that is of high importance regarding RSI. However, the time history of the net head during load rejection did not agree with the behavior of the experimental data at a satisfactory level. The percentage deviations given for both cases in Figure 10 show that all results are still within 6% of the comparison data.

Figure 8. Net head (left) and VL2 pressure (right) during transient load operation at BEP2HL (Case 9).

IOP Conf. Series: Journal of Physics: Conf. Series **782** (2017) 012002 doi:10.1088/1742-6596/782/1/012002

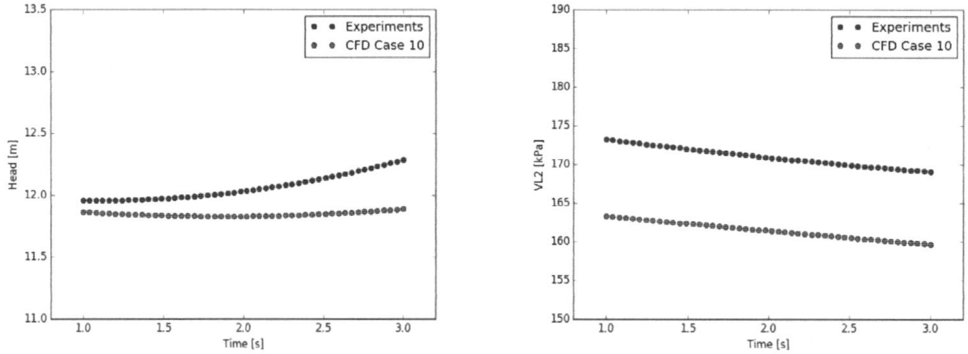

Figure 9. Net head (left) and VL2 pressure (right) during transient load operation at BEP2PL (Case 10).

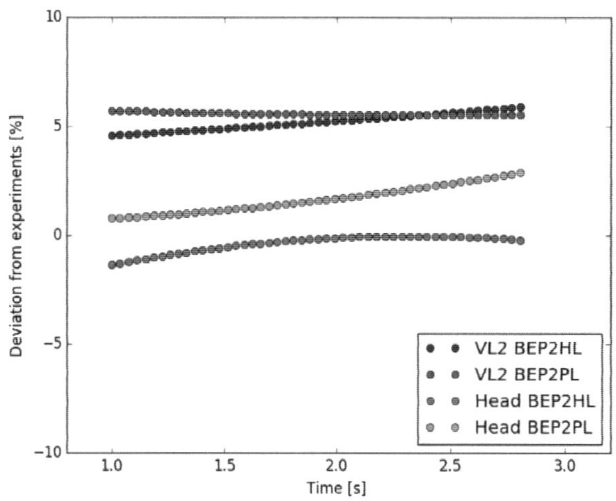

Figure 10. Percentage deviation from experiments (Case 9 & 10).

Illustrated by streamlines initiated at the runner/draft tube interface, Figure 11 shows the vortex rope in draft tube after end of guide vane rotation for both load rejection and load acceptance. While there were no flow circulations of significance in the draft tube at BEP (not shown), strong swirls of opposite direction were generated during opening and closing of the guide vanes. These are known effects in the literature when considering stationary load conditions [2]. The BEP2HL case shows a more defined swirl with higher fluid velocities in accordance with the higher mass flow loading.

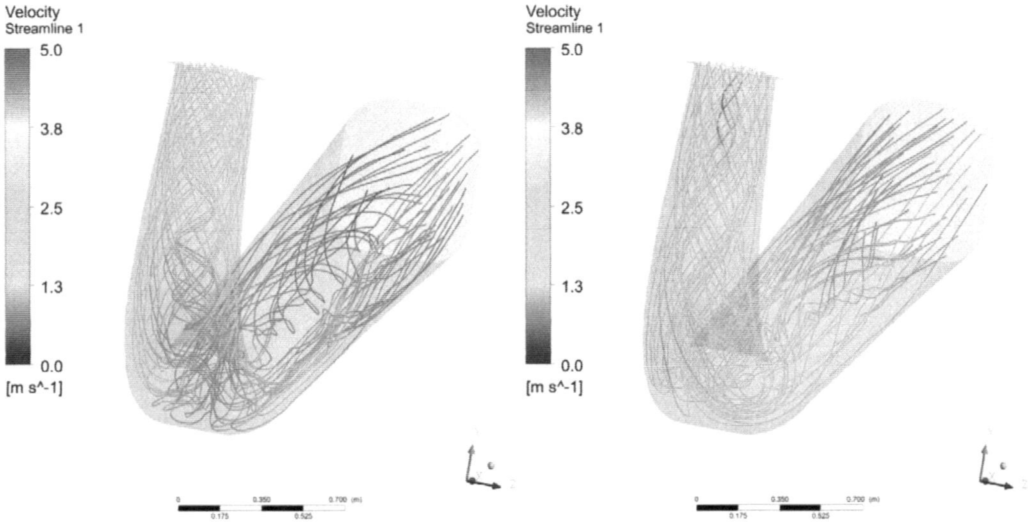

Figure 11. Streamlines from runner/draft tube interface at end of guide vane rotation. Left: BEP2PL, right: BEP2HL.

Representing flow field in blade passage, Figure 12 and Figure 13 give the velocity in rotating frame and the velocity in stationary frame, respectively. The contours have been visualized on an isosurface going through the center of the runner and following its out-of-plane direction. The absolute pressure in the runner is seen in Figure 14, clearly experiencing the familiar RSI pressure behavior at the circumference of the domain [2].

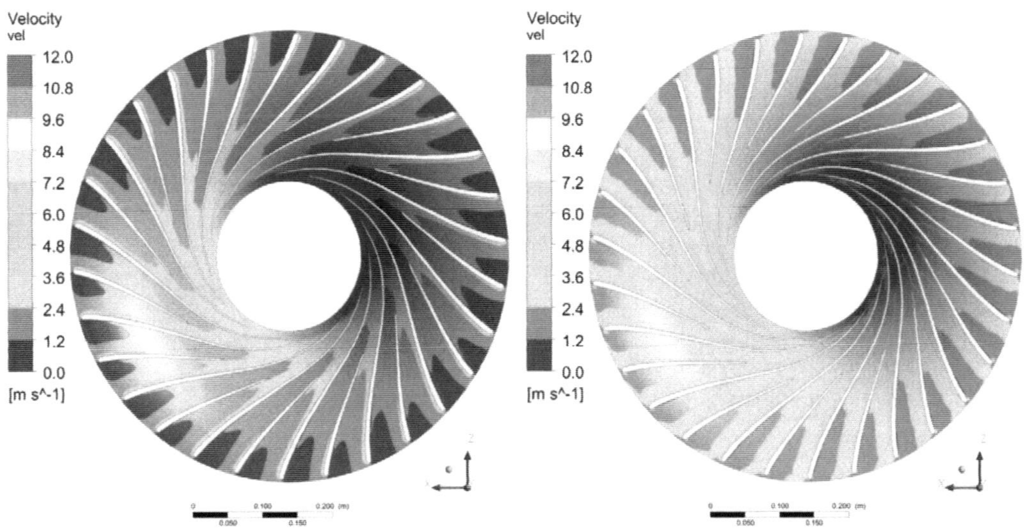

Figure 12. Velocity in rotating frame at end of guide vane rotation. Left: BEP2PL, right: BEP2HL.

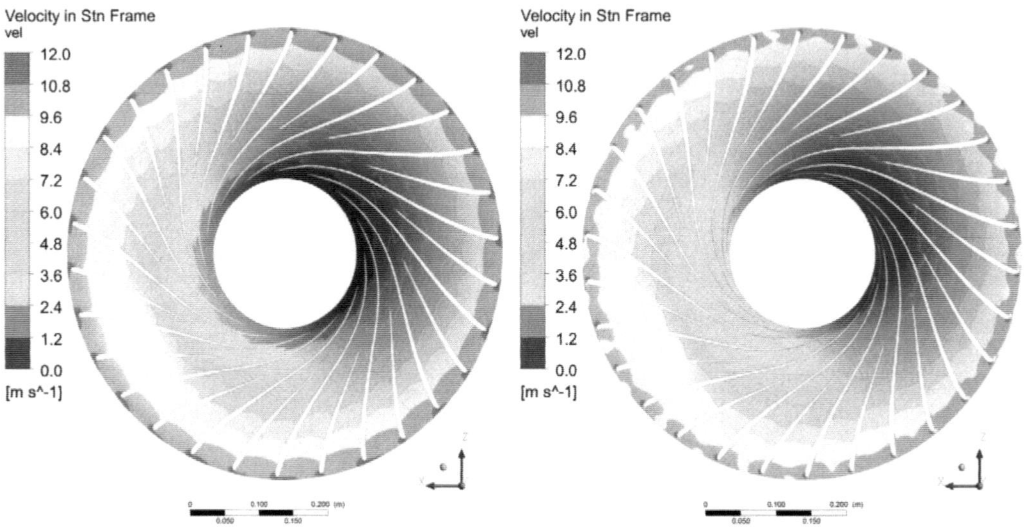

Figure 13. Velocity in stationary frame at end of guide vane rotation. Left: BEP2PL, right: BEP2HL.

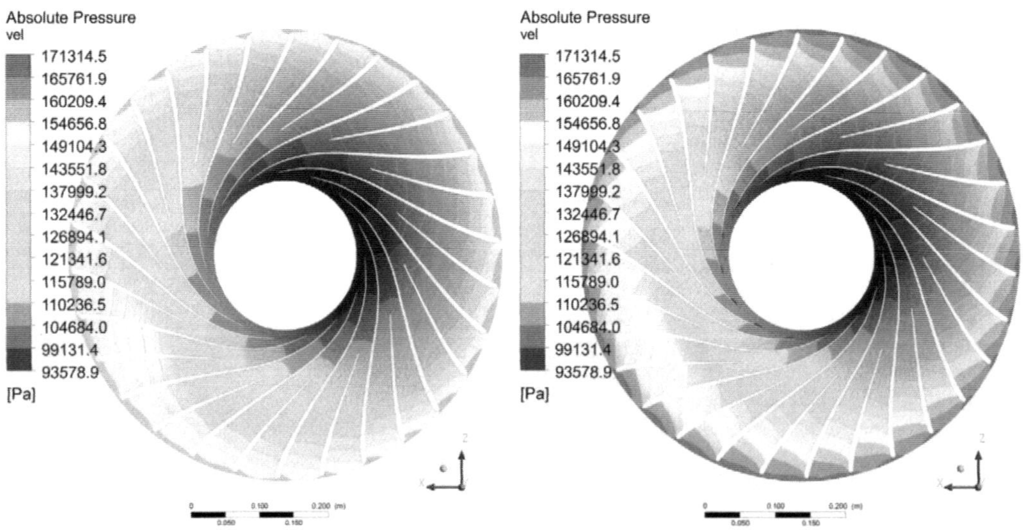

Figure 14. Absolute pressure at end of guide vane rotation. Left: BEP2PL, right: BEP2HL.

5. Concluding remarks

Steady-state and transient CFD simulations have been performed on the full 360° Francis-99 hydraulic turbine, including load acceptance and rejection utilizing mesh deformation. The work was performed for the second Francis-99 workshop and as part of the HiFrancis project and the results have been compared to the existing measurements performed at the Water Power Lab at NTNU.

The main findings from the mesh dependency study and comparison with experimental data were as follows:

- A mesh independent solution was reached for a mesh with approximately 42 million nodes and near wall y+ values in the order of 3 in the spiral casing and runner. Tetrahedral elements were used in the spiral casing.
- The comparison between torque and torque based on ΔUCu proved that the velocity field was captured at sufficient accuracy as the difference between the two was low (<1 %) for the mesh independent solution.
- Average values for global variables such as net head, torque, inlet pressure, outlet pressure and hydraulic efficiency were in general predicted very well. The results improved further by including a transient rotor. Pressure amplitudes have not been subject to investigation in the present paper.

During load acceptance and rejection, the guide vane rotation was performed in accordance with the operating scheme reported in [1]. The transient results obtained at the mesh independent solution was used as initial condition with the main findings as follows:

- In general, good agreement was found when comparing the trends of the global measurement variables with results being within 6% off from the experiments. More specifically and important for capturing the effects of RSI, the pressure in vaneless space (VL2) agreed very well with the experiments.
- The initial deviations on absolute values were in accordance with the results seen for the stationary load cases used as starting conditions. However, net head during load rejection did not follow the experimental values at a satisfactory level.

The observation made for the net head at load rejection will be investigated further, including alterations such as applying a pressure condition at the volute inlet rather than explicitly specifying the mass flow. The latter will be in closer resemblance with both the experiments as well as real life situations. Other known deviations from the experiments are that the simulations were performed with one single upstream pressure probe, while the experimental data reports an average of four circumferentially placed probes at that location. Also, updated guide vane position from the experiments shows a deviation of about 0.02 degrees.

At the time of writing, further investigations are currently ongoing as part of the HiFrancis project.

References

[1] NTNU, "Francis-99 Second Workshop," [Online]. Available: https://www.ntnu.edu/nvks/test-case.

[2] C. Trivedi, M. J. Cervantes, B. K. Gandhi and O. G. Dahlhaug, "Experimental and Numerical Studies for a High Head Francis Turbine at Several Operating Points," *Journal of Fluids Engineering,* vol. 135, 2013.

[3] C. Trivedi, M. J. Cervantes and O. G. Dahlhaug, "Numerical Techniques Applied to Hydraulic Turbines: A Perspective Review," *Applied Mechanics Reviews,* vol. 68, 2016.

[4] C. Trivedi, *NTNU, Email and phone correspondance throughout project.*

[5] J. Schiffer, H. Benigni and H. Jaberg, "Analysis of the leakage behaviour of Francis turbines and its impact on the hydraulic efficiency - A validation of an analytical model based on CFD-results," *Jornal of Fluids Engineering,* 2016.

[6] K. Kloster, "Pressure pulsation inside a Francis turbine runner (Master Thesis)," NTNU, 2016.

Simulation of the Francis-99 Hydro Turbine During Steady and Transient Operation

Yuvraj Dewan
Senior Application Engineer, CD-adapco. 200 Shepherds Bush Road, London W6 7NL, UK
E-mail: yuvraj.dewan@cd-adapco.com

Chad Custer
Technical Specialist: Turbomachinery, CD-adapco. 21800 Haggerty Road, Northville, MI-48335, USA
E-mail: chad.custer@cd-adapco.com

Artem Ivashchenko
Project Engineer, Mechanical Solutions Inc. 11 Apollo Drive, Whippany, NJ 07981, USA
E-mail: artem.ivashchenko@mechsol.com

Abstract. Numerical simulation of the Francis-99 hydroturbine with correlation to experimental measurements are presented. Steady operation of the hydroturbine is analyzed at three operating conditions: the best efficiency point (BEP), high load (HL), and part load (PL). It is shown that global quantities such as net head, discharge and efficiency are well predicted. Additionally, time-averaged velocity predictions compare well with PIV measurements obtained in the draft tube immediately downstream of the runner. Differences in vortex rope structure between operating points are discussed.

Unsteady operation of the hydroturbine from BEP to HL and from BEP to PL are modeled. It is shown that simulation methods used to model the steady operation produce predictions that correlate well with experiment for transient operation.

Time-domain unsteady simulation is used for both steady and unsteady operation. The full-fidelity geometry including all components is meshed using an unstructured polyhedral mesh with body-fitted prism layers. Guide vane rotation for transient operation is imposed using fully-conservative, computationally efficient mesh morphing. The commercial solver STAR-CCM+ is used for all portions of the analysis including meshing, solving and post-processing.

1. Introduction

According to the 2016 Global Status Report on renewable energy, in 2015 over 23% of global power production came from renewable energy sources, of which over 57% were hydropower sources [1]. Most of that capacity belongs to large hydroelectric plants using Francis turbines to generate that power. Furthermore, some countries, such as Iceland, Brazil, Canada and Norway, derive a significant majority of their electric power from hydroelectric plants. As a result, smooth and continuous operation of Francis turbines and a thorough understanding of the fluid flow inside a turbine at best efficiency point and off-design conditions is important for hydropower producers.

Advancements in computational fluid dynamics (CFD) have contributed greatly to component design [2, 3] and performance prediction of Francis turbines. This paper focuses on the use of CFD for performance prediction and detailed characterization of the flow-field. The experimental processes involved in performance prediction of the complete turbine system from scroll to draft tube are expensive, time-consuming and demanding in terms of manpower. By comparison, a CFD analysis of the turbine system is faster and, once a model is created, can produce additional results with very little effort. Still, due to constraints on computational power, many simulations have been performed with coarse meshes and steady state methods. With this method, moving components of the turbine are modeled with moving reference frames (MRF) instead of physically rotating the runner mesh relative to the mesh of stationary components [4, 5, 6, 7]. While undoubtedly useful for clarification of a wide variety of issues, such analyses fail to capture important transient events, such as vortex rope development, wake interactions and pressure oscillations, especially at off-design conditions. Additionally, these unsteady effects influence global, time-averaged quantities, which limits the accuracy of steady simulation methods. Thankfully, continuous advancements in both commercial CFD codes and, more importantly, in processing power and its cost have made mesh refinement and unsteady rigid body motion (RBM) possible for a wider audience of engineers and researchers [8, 9, 10].

This paper presents the results of unsteady CFD analysis of the Francis-99 test case. Three operating conditions, including two off-design conditions, are simulated. The turbine was modeled from the inlet of the scroll to the discharge of the draft tube into the tank, using upwards of 15 million cells. Analysis was conducted using the commercial code STAR-CCM+ produced by CD-adapco [11]. The results are compared against the data gathered by the researchers at the Norwegian Hydropower Center (NVKS).

2. Reference Experiment

The Francis-99 hydroturbine is an experimental test rig housed at the Norwegian Hydropower Center (NVKS) at the Norwegian University of Science and Technology (NTNU). The experiment contains a scaled down turbine which is based on the characteristics of the turbines found at the Tokke hydroelectric powerstation in Telemark, Norway. It has been scaled down from the actual operation scale to lab scale. Numerous experimental investigations have been performed including the characterization of steady operation at numerous operating conditions, load acceptance and rejection, emergency shutdown, total load rejection, and runaway.

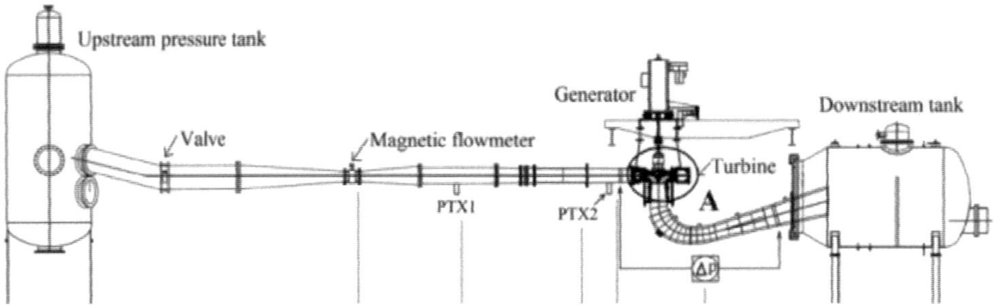

Figure 1: Francis-99 experimental setup schematic [12]

Geometric CAD data, operating conditions and experimental measurements are provided online for steady operation at BEP, PL and HL [12]. Additionally, transient operation for BEP to HL and BEP to PL conditions are characterized. This open access has facilitated numerous numerical investigations that are part of the continuing Francis-99 workshop series. The main global quantities of interest such as net head, discharge and hydraulic efficiency are measured for both steady and unsteady operation. Additionally, velocity measurements in the draft tube just downstream of the runner are obtained using PIV for steady operation conditions as indicated by the solid lines in Figure 2. Simulation data will be correlated to these experimental data.

3. Simulation Setup

3.1. Geometry

The geometry for the Francis-99 turbine is openly provided by the Francis-99 committee in a variety of CAD and mesh formats [12]. For this study, the STEP format of the geometry is used. The simulation domain consists of 4 main regions: spiral casing, guide vanes, runner and draft tube. The entirety of the simulation domain is shown in Figure 2a and a view of the runner is shown in Figure 2b.

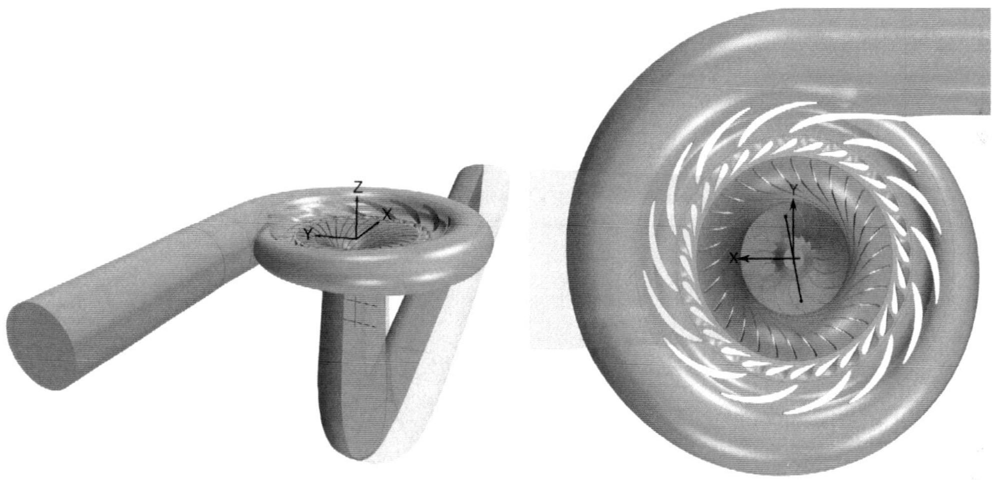

(a) Simulation domain (b) Geometry viewed from draft tube

Figure 2: Francis-99 CAD geometry shown with the global coordinate system and PIV measurement line locations

No CAD defeaturing is performed. All geometric details including fillets, chamfers, and small edges provided in the CAD are retained in the model. No secondary flows, such as the seals between runner and casing, were provided and so are not included in the simulation domain.

3.2. Computational Mesh

A polyhedral mesh with prism layers is generated for the entire domain. A polyhedral cell is an unstructured mesh topology where each cell can have an arbitrary number of faces. This increases cell connectivity to neighbors and increases the degrees of freedom available to the mesher. As a result, higher quality cells can be obtained with low skewness and smooth volumetric growth. Consequently, the polyhedral mesh technology generally shows better convergence than a tetrahedral mesh for a given number of cells [13].

In addition to the numerical benefits of the polyhedral cell, the topology allows for better modeling of the physics and geometry of the system. The unstructured nature of the mesh allows for arbitrarily complex geometry to be modeled without the need for geometric simplification. The polyhedral mesh is well suited to flows in which there is no clear flow direction, such as the vortex rope. With many more faces than a structured or tetrahedral mesh, faces with the largest face flux are generally closer to normal to the velocity field, resulting in better stability and lower error. Figure 3b shows the uniform and direction-independent nature of the mesh in the draft tube immediately downstream of the runner.

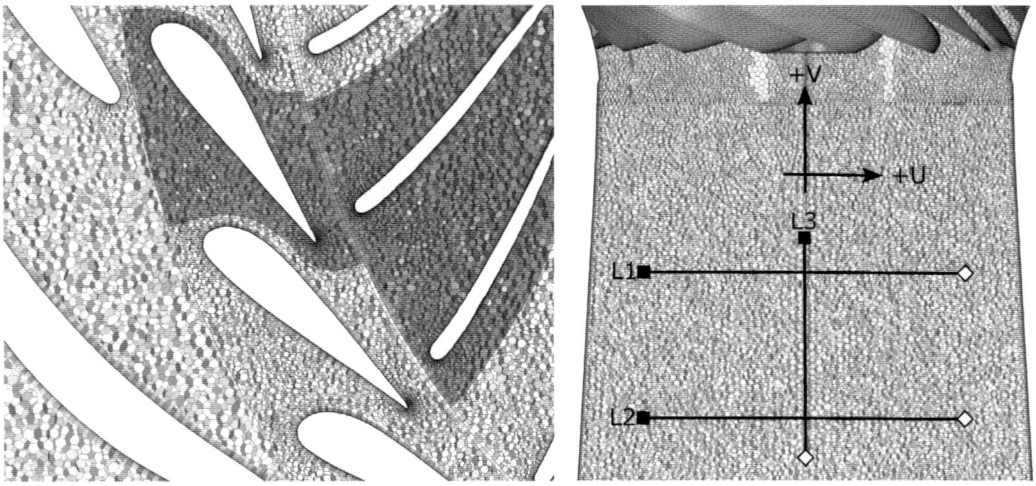

(a) Midspan mesh of bladed components showing duplicated periodic sector

(b) Draft tube mesh downstream of runner with velocity measurement locations and directions

Figure 3: Polyhedral mesh with prism layers used to discretize domain

A layer of thin, anisotripic cells, known as prism layers, are grown outward from all wall surfaces. These prism layers accurately capture the large gradients in the flow-field approaching the no-slip condition imposed at the wall. The prism layer mesh at mid-span of the bladed components is shown in Figure 3a. Ten prism layers are used at all wall surfaces. The specified first cell thickness is set such that the wall y+ is generally less than five. For the BEP operating point, the average y+ for all wall boundaries in the domain is 4.5. In addition to the prism layers used to capture the wall gradients, one thick prism layer is used at the interface between all rotating and stationary regions. This ensures cell orthogonality and consistent sizing at the interface.

To maintain self-similarity, each periodic and symmetry section is meshed once, then repeated. As shown in Figure 3a, only one guide vane and runner flow passage is meshed, then the resulting grids are duplicated, patterned and conformally fused into a unified domain. Likewise, one half of the spiral casing and draft tube is meshed and mirrored about the plane of symmetry and fused. Approximately 14M cells are used to model the complete domain. The bladed components represent the largest cell count with approximately 4M cells used to mesh the guide vanes and 5M cells used to mesh the runner. The spiral casing and draft tube consists of 3.5M cells and 1.3M cells, respectively.

3.3. Boundary Conditions

3.3.1. Steady Operation From the experimental measurements, both the flow rate and pressure at the spiral casing inlet are observed to be unsteady. The unsteadiness observed at this location is due to the dynamics of the system including the blade passing effects. The phase lock between the observed oscillations at the measurement locations relative to the runner position is not available. For these reasons, constant values are imposed as boundary conditions for steady operation. A consequence of specifying constant values for boundary conditions is that some of the observed oscillations will be artificially damped. To accurately model the oscillations observed at the measurement locations, the domain must be extended to a secondary station where values are known to be constant. This is neither feasible, or necessary. The effect of imposing constant values as boundary conditions for the steady operation case does not greatly influence the quantities of interest.

With it established that constant values will be used for the boundary conditions of the steady operation cases, the type of boundary condition imposed at the inlet and outlet must be determined. The measurement of static pressure is often more reliable than that of flow rate. Additionally, the static pressure at the inlet of a Francis turbine is usually well known based on the vertical elevation change from the inlet of the turbine to free surface of the water powering the turbine. For these reasons, a pressure condition will be used as the inlet condition. Total pressure is specified as the inlet boundary condition as given by equation 1.

$$P_{t1} = P_1 + \frac{1}{2}\rho\left(\frac{Q}{A_1}\right)^2 \tag{1}$$

With total pressure fixed at the inlet, the outlet boundary condition can be imposed as either a constant static pressure, or discharge. Again, it is common to have reliable data for the outlet static pressure of a Francis turbine. Additionally, the relative pressure difference from inlet to outlet is strongly linked to net head, which greatly influences performance. For these reasons a constant value of static pressure is imposed at the domain outlet. All other required boundary values are extrapolated from the interior of the domain. The test rig measures the outlet static pressure using a ring manifold as specified in the IEC 60193 standard [14]. However, this pressure is measured at a section upstream of the domain, not at the location where boundary conditions will be specified. Consequently, the measured pressure must be adjusted before it is applied as a boundary condition. The static pressure applied at the downstream boundary is approximated using Bernoulli's principle as given in equation 2. Let *1* be the measurement plane and *2* be the domain outlet.

$$P_1 + \frac{1}{2}\rho v_1^2 + \rho g z_1 = P_2 + \frac{1}{2}\rho v_2^2 + \rho g z_2 \tag{2}$$

The velocity at each section can be estimated using the continuity equation. The area for each location is measured in STAR-CCM+ and the value for discharge is taken from the operating condition.

$$Q = vA \tag{3}$$

The absolute static pressure at the outlet can be calculated by combining equations 2 and 3.

$$P_2 = P_1 + \rho\left[\frac{Q^2}{2}\left(\frac{1}{A_1^2} - \frac{1}{A_2^2}\right) + g\left(z_1 - z_2\right)\right] \tag{4}$$

The provided value for gravitational acceleration (g) and density (ρ) are 9.82 m/s^2 and 999.8 kg/m^3, respectively. The difference in the vertical positions ($z_1 - z_2$) is measured from the CAD to be 0.4031 m. The cross sectional areas A_1 and A_2 are measured from the CAD to be 0.2292 m^2 and 0.3294 m^2, respectively. The discharge (Q) and measured outlet pressure (P_1) are provided from the experimental data. While running the simulations, it was confirmed that the outlet pressure measured in the experiment matched well with the pressure at the measurement plane after the outlet corrections.

3.3.2. Transient Operation

Inlet For the transient operating conditions, the inlet pressure is specified as a function of time. The inlet pressure profile provided from the experiment must be processed before it is applied to the simulation. The inlet pressure data provided contains high frequency content related to the blade passing frequency and sampling noise. This high frequency content is filtered using seasonal decomposition with moving averages [15].

An additive model is used for the seasonal decomposition, meaning that the original signal is divided into the following components: trend, seasonal and error. The sum of these components returns the original signal. The trend component is first calculated using a symmetric moving average. This provides the long-term or low frequency trends in the data. The original signal is then 'de-trended' by subtracting the trend component. The 'de-trended' data is then corrected to a zero mean to give the seasonal component. This component provides the high frequency content from the original data. Finally, the error component is calculated by taking the difference between the original signal and the sum of the trend and seasonal components. This error component represents the noise and turbulence in the data. Figure 4 shows the decomposition of inlet pressure at the load acceptance transient operating condition (BEP to HL). The trend component is the boundary condition that is applied for simulation.

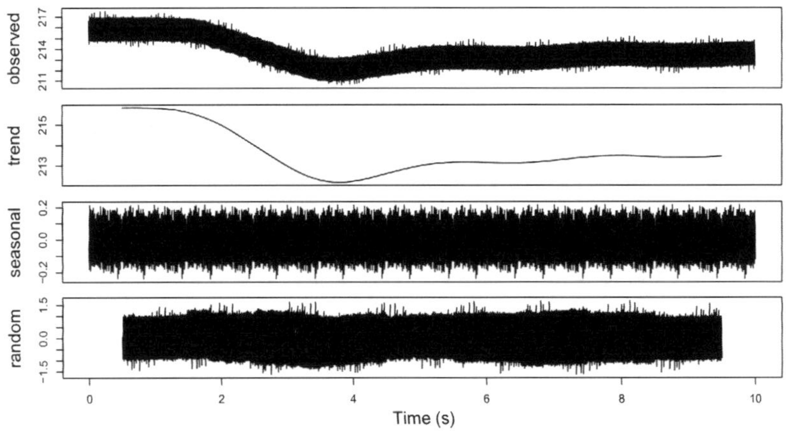

Figure 4: Time series decomposition of the inlet pressure for the load acceptance condition. Units for all y axes are kPa

Outlet Measured static pressure at the outlet measurement plane is not directly provided from the experiment for the transient operation cases. The net head (H), discharge (Q) and inlet static pressure (P_1) are provided as functions of time for the transient operation case. The static pressure at the measurement plane is calculated from the provided quantities using the definition of net head as given in equation 5 and the relation between discharge, area and velocity given in equation 3.

$$H = \frac{P_1 - P_2}{\rho g} + \frac{v_1^2 - v_2^2}{2g} + z \tag{5}$$

This calculated static pressure is then corrected for application at the boundary condition location as given in equation 4.

IOP Conf. Series: Journal of Physics: Conf. Series **782** (2017) 012001 doi:10.1088/1742-6596/782/1/012001

3.4. Solver Setup

A Semi-Implicit Method for Pressure-Linked Equations (SIMPLE) algorithm is used to solve the incompressible Reynolds-Averaged Navier-Stokes equations. Spatial gradients are computed to second-order accuracy and a second-order, implicit unsteady method is used to advance time. Rotation is modeled using rigid body motion (RBM) with in-place, direct interfaces between rotating and stationary regions that are recomputed at each time-step. A time-step corresponding to one degree of rotation is used and twenty inner iterations are used to converge each time-step. Menter's SST $k-\omega$ turbulence model is used with an all y+ implementation. However, as discussed in Section 3.2, the mesh was generated so that a low y+ treatment is used on the vast majority of wall boundaries.

3.5. Guide Vane Motion

For transient operating conditions, each guide vane rotates about its own axis as a function of time defined by a piece-wise linear rotation profile. This motion is modeled by a mesh morphing approach. The walls of each guide vane are set to rigidly rotate as specified by the transient operation of the turbine. Hub and shroud boundaries of the guide vane region are set to a 'floating' condition which allows the mesh on these boundaries to freely adjust to the motion applied to the other boundaries. Inlet and outlet interface boundaries to the guide vanes are fixed. This method is fully conservative and computationally efficient. Remeshing of the domain is not required and there is no interpolation of the solution from one time-step to the next. Figure 5 shows the starting and final guide vane position for the load reduction condition.

Another approach that could be implemented for guide vane motion is the Overset or Chimera grid approach [16]. In this technique, every component which has relative motion to another is meshed independently. This leads to a series of overlapping meshes. At each time-step, all of the overlapping grids are combined into a single grid on which the physical equations are solved. The details of this mesh combination is outside the scope of this report. For this study, each guide vane would require its own Overset (or overlapping) region. This means that each guide vane region needs an interface with the background region and two of its neighboring guide vanes, which would result a total of 57 overset interfaces that need to be resolved at each time step. There is a high computational expense associated with the computation of a large number of interfaces, and it is the reason that the morphing approach is preferred for this study.

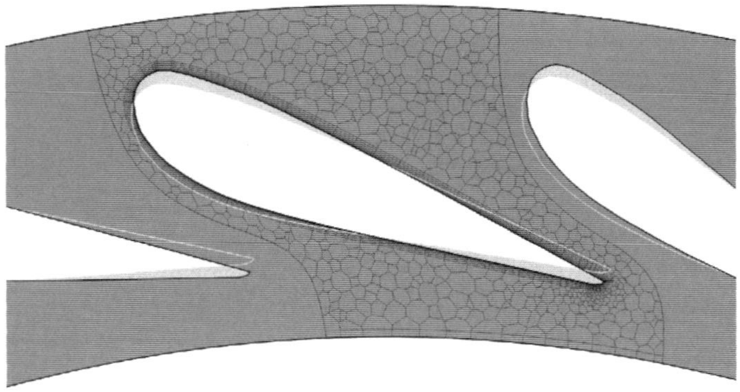

Figure 5: Guide vane position at BEP (grey mesh) and PL (cream).

4. Results

4.1. Steady Operation

A simulation was run for each of the three operating conditions that are documented in the reference experiment (PL, BEP and HL). The simulation was run long enough for integral quantities to reach cycle steady values. The amount of time taken for each operating condition to reach cycle steady behavior is variable. The PL case exhibits certain low frequency trends, which requires the solution to be run over a longer period of time. The HL case requires a shorter run time to reach convergence, and the BEP case even shorter.

Figure 6 shows the predicted values of the integral quantities as compared to the reference experiment. The simulation predictions for all of the relevant quantities of interest agree well with experiment across all three steady operating conditions. Table 1 shows a summary of the percent error for each quantity as compared to the experiment. The largest deviation from experimental results is the over-prediction of hydraulic efficiency for all three operating conditions. This is largely due to the exclusion of all secondary flow passages from the simulation. In the physical test, secondary flow passages such as the labyrinth seal introduce losses into the system. Omitting these passages from the simulation is a primary cause for the over-prediction of efficiency. It is important to note that the inclusion of these passages would further reduce the torque and discharge predictions.

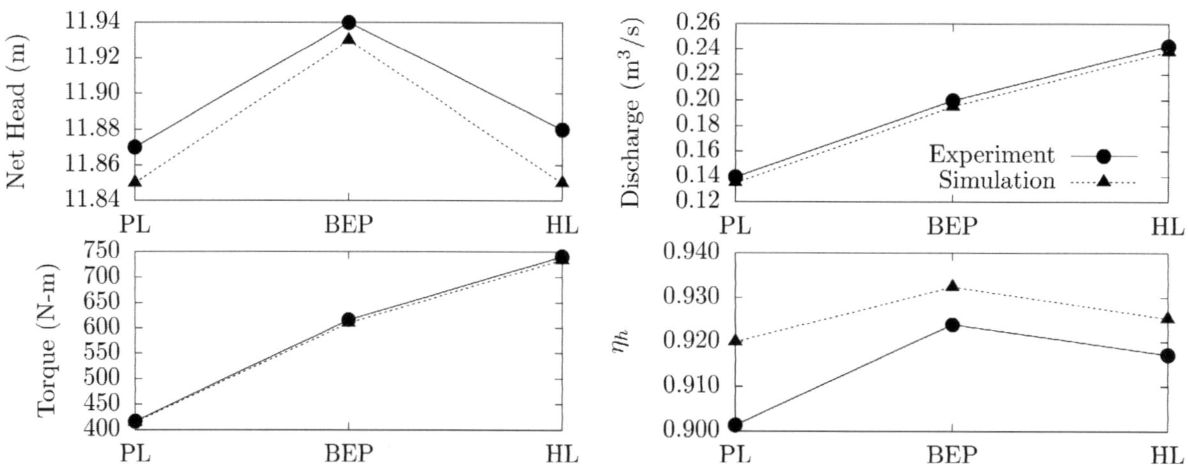

Figure 6: Predictions of quantities of interest for steady state operation

Table 1: Percent error of predicted integral quantities against reference experiment

Condition	Net Head	Discharge	Torque	Hydraulic Efficiency
PL	-0.17%	-3.31%	-0.33%	2.07%
BEP	-0.08%	-2.30%	-0.99%	0.92%
HL	-0.25%	-1.84%	-0.88%	0.89%

The reference experiment also provides velocity data along three lines in the draft tube. The locations of these sample lines are shown in Figure 3b. Velocity is measured in a coordinate system that is centered with the runner axis and oriented as shown in the same figure. The probes are oriented in two different directions to provide two distinct pieces of information. As shown in Figure 3b, measurement lines L1 and L2 are oriented transverse to the draft tube, while L3 is aligned with the runner axis. The velocity profiles along L1 and L2 provide information on the size and strength of the vortex rope. Probe L3 characterizes how the strength of the vortex rope changes moving downstream from the runner. For the remainder of the results, the probe locations will be presented as a function of normalized position between 0 and 1. A normalized location of 0 is shown as a black square and a normalized location of 1 is shown as a white diamond in Figure 3b.

An investigation of the experimentally measured data is important to understand the length of the sample required for obtaining a mean flow-field. The measurements of velocity in the draft tube in the reference experiment show that there is a significant difference in frequency of oscillations for each condition. For averaging the simulation, we must at least run for a time equal to the period of the largest wavelength of oscillation from the experiment. This will ensure that flow features across all the relevant time-scales have been captured. The data shows that the lowest frequency of velocity measurement is shown in the PL case. This low frequency structure has a time-period of three runner revolutions so all velocity probe data is averaged over this period after reaching cycle steady convergence.

Comparing simulation data to the measured average velocities provides an indication of how well the flow features are predicted within the draft tube. The averaged velocity measurements from the experiment and simulation for the PL steady operating condition are shown in Figure 7. The predicted U velocity component is quite close to zero for all three measurement lines, which means that the flow is not strongly biased towards this Cartesian direction. The V velocity component provides some information about the shape of the vortex rope structure. The V velocity component trends show a large area of low velocity towards the center of the draft tube. This suggests a the presence of a large, unsteady vortex rope. Figure 10a illustrates the large size of the vortex core. Figure 12a also shows the existence of a large and unstructured vortex rope by presenting the time-averaged streamlines of the vortex rope. Predictions along L3 for V also highlight the quick reduction in the intensity of this vortex rope. Data for both velocity components correlates well with experimental measurements.

The averaged velocity probe measurements from the experiment and simulation for the BEP steady operating condition are shown in Figure 8. As was the case with the PL operating condition, the U velocity components for this case are close to zero for all three lines. Studying the V velocity trends shows a very narrow region of relatively low velocities near the center of the draft tube. This in turn suggests the existence of a narrow or 'tight' vortex rope. This is confirmed by the time averaged V velocity component contours shown in Figure 11b and the time-averaged streamlines of the vortex rope shown in Figure 12b. Probe L3 predictions for V show a weakening of this vortex as one moves downstream in the draft tube. In general, the CFD predictions correlate well with experiment, with the largest deviation being the under-prediction of the velocity magnitude in the axial direction.

The averaged velocity probe measurements from the experiment and simulation for the HL steady operating condition are shown in Figure 9. The simulation predicts a close to zero U velocity for all lines again, but the experiment suggests that there is some directionality for U for this operating condition. The experimental measurements show that the U velocity is slightly negative for each of the probes. The V velocity component predictions show the existence of a very strong vortex rope for this case. With both L1 and L2 probes, V almost reaches complete stagnation at the center of the draft tube. Compared to the BEP case, the vortex rope also appears to be slightly wider. This is clearly evidenced by the time averaged V velocity contours shown in Figure 11c and the time-averaged vortex rope streamlines shown in Figure 12c. The simulation predictions for the V velocity component match the experimental measurements quite

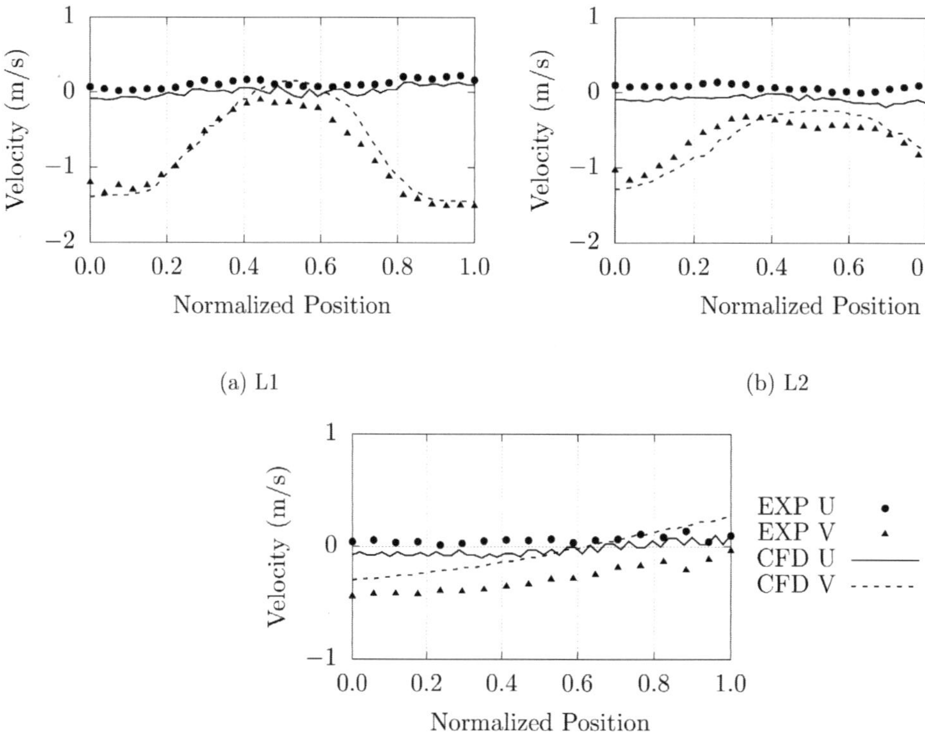

(a) L1 (b) L2

(c) L3

Figure 7: Simulation predictions and experimental results for averaged velocities in the draft tube for the PL operating condition

well.

4.2. Unsteady Operation

The unsteady operation of the turbine is the transient phenomenon in which the turbine changes between steady points of operation. The transient conditions studied in this study are: 1. Load acceptance (BEP to HL), and 2. Load reduction (BEP to PL). As mentioned in section 3.3, the inlet and outlet pressures are calculated as a temporal profile from the experimental data. The extracted quantities of interest are net head and discharge. The discharge measured in the experiment for both of these operating conditions is not accurate [12]. The experimentalists recommend to use a linear profile between the measurements for the two steady operating conditions. It is also important to note that the experimental reference suggests that the measurements of the key quantities may not be synced temporally with the guide vane rotation, so some artificial lead or lag is expected in the response.

Comparisons of the experimental and simulation measurements are shown in Figures 13 and 14 respectively. The simulation net head results follow the experimental measurements quite closely. The simulation prediction shows an abrupt change at the exact time when the guide vanes start their rotation (time = 1 sec). This abrupt change is not found in the experiment. This can be explained by the fact that the simulation is run with a constant density approach,

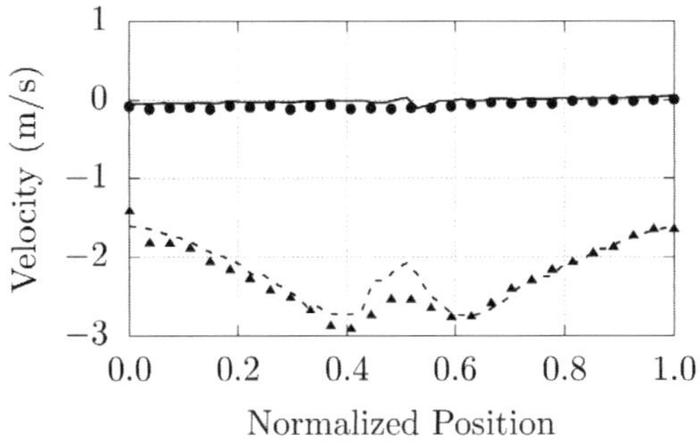

(a) L1

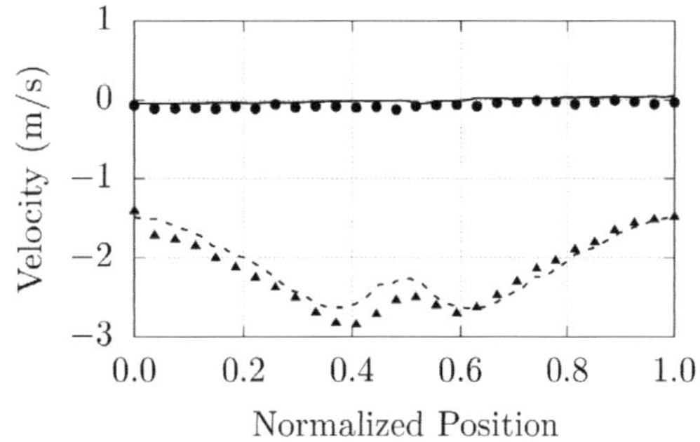

(b) L2

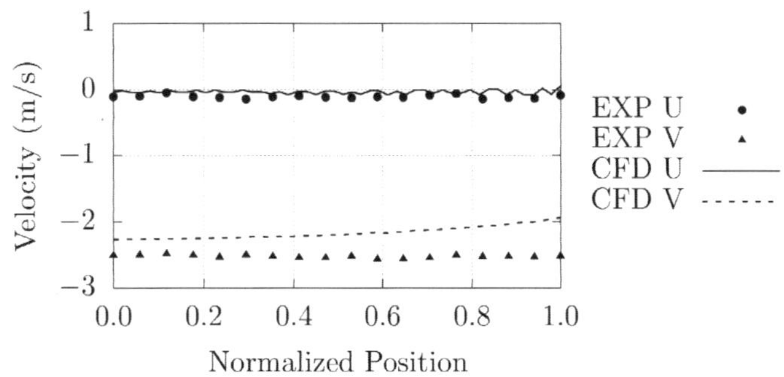

(c) L3

Figure 8: Simulation predictions and experimental results for averaged velocities in the draft tube for the BEP operating condition

therefore any discontinuity in simulation conditions causes an abrupt change everywhere in the domain. Adding compressibility to the modeling approach would help to absorb some of these effects (similar to actual operation). The amplitude of oscillations of net head is also under-predicted, but this is to be expected as explained in section 3.3. The discharge values show a linear profile during guide vane rotation. This follows the expected trend exactly as given by the experimentalists. The quantitative simulation measurements also track the experiment well. The slope of the linear profile is completely dependent on the start and final predictions of the simulation, i.e. a different slope is indicative of the steady operating conditions at the start and end of transient operation not being captured exactly. An estimate of discharge prediction error for steady operating conditions can be found in section 4.1.

5. Conclusions

Numerical predictions of the Francis-99 hydroturbine using the commercial CFD code STAR-CCM+ correlate well with experiment. All simulations use implicit unsteady time integration with rigid body motion to model runner rotation. An unstructured polyhedral mesh is able to accurately capture all geometric features and is well suited to modeling complex flow structures such as the vortex rope. Prism layer cells are used to resolve wall gradients, which is important to accurately predicting runner torque and, therefore, efficiency.

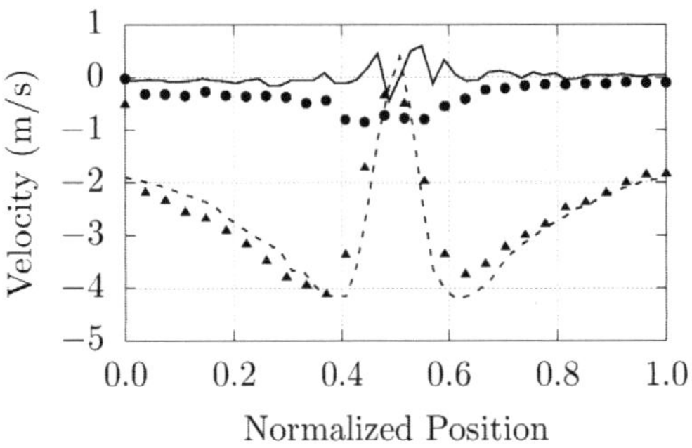

(a) L1

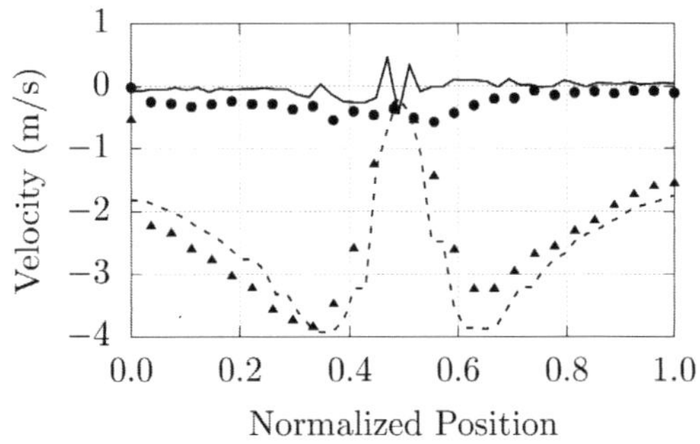

(b) L2

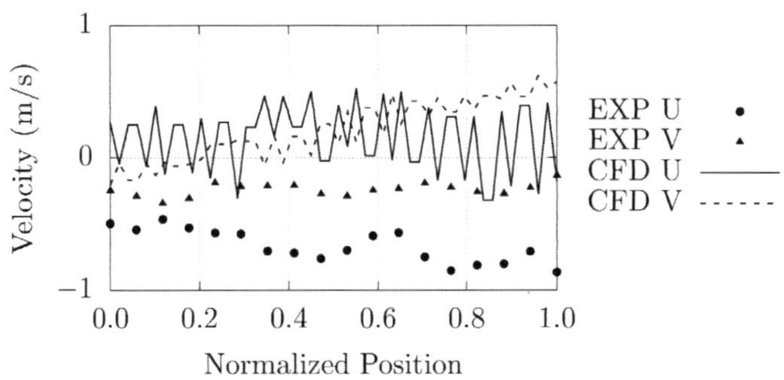

(c) L3

Figure 9: Simulation predictions and experimental results for averaged velocities in the draft tube for the HL operating condition

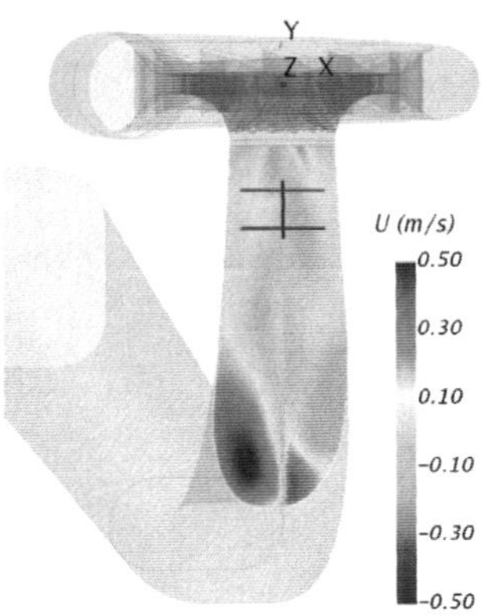

(a) PL

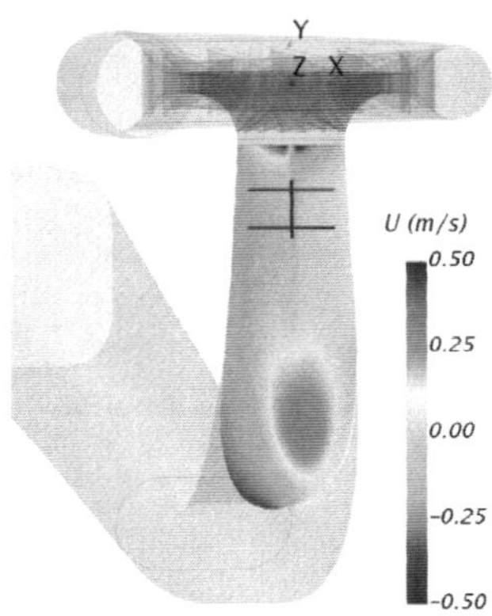

(b) BEP

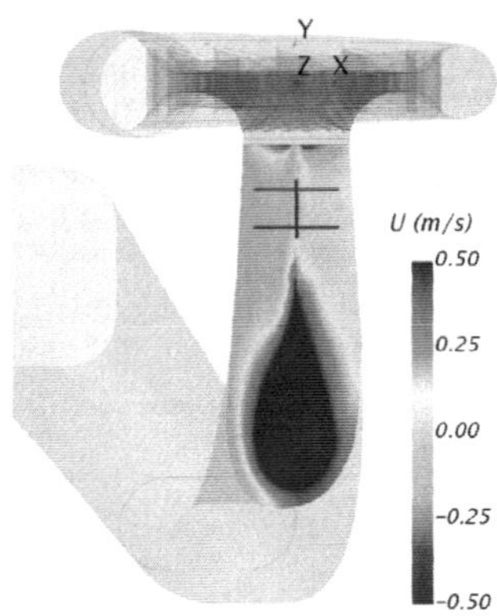

(c) HL

Figure 10: Time averaged U velocity contours for the steady operating conditions. The coordinate system shown is used for the velocity measurements.

IOP Conf. Series: Journal of Physics: Conf. Series **782** (2017) 012001 doi:10.1088/1742-6596/782/1/012001

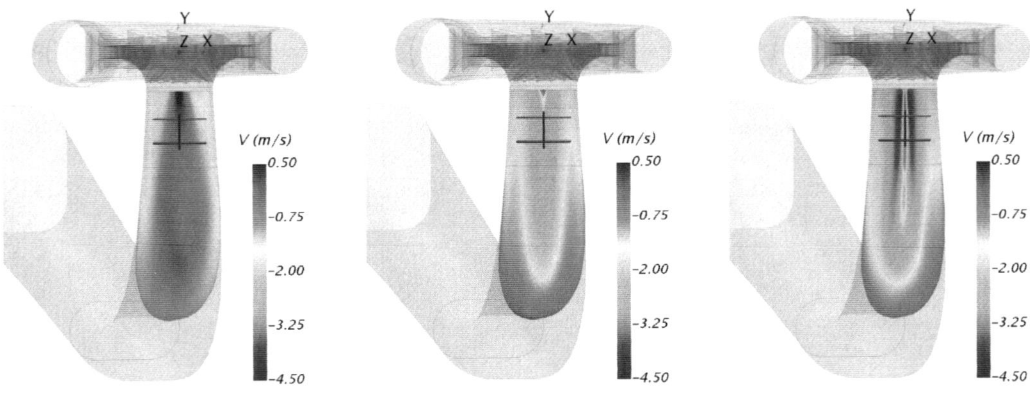

(a) PL (b) BEP (c) HL

Figure 11: Time averaged V velocity contours for the steady operating conditions. The coordinate system shown is used for the velocity measurements.

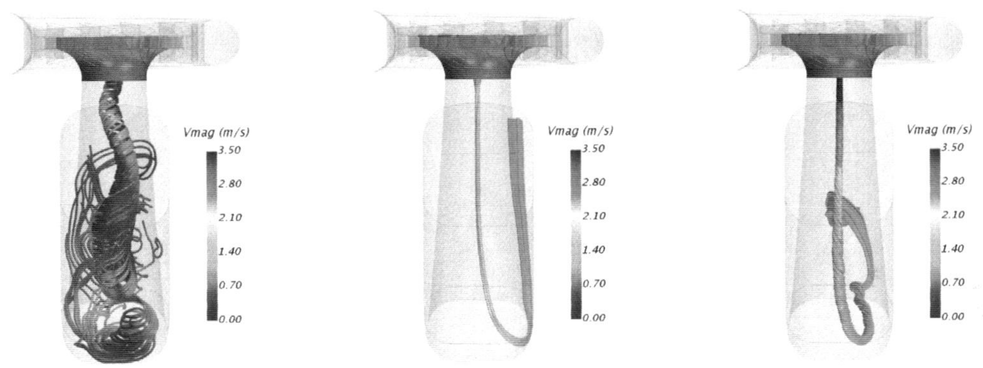

(a) PL (b) BEP (c) HL

Figure 12: Time averaged streamlines of the vortex rope for the steady operating conditions

Steady operation was analyzed at BEP, PL and HL conditions. Net head, discharge, runner torque and hydraulic efficiency all correlate closely with experimentally observed values. Nearly all quantities are predicted within 3.5% of the experimental observation. Part load operation shows the greatest discrepancy. This operation condition contains the most complex flow features with many unsteady time-scales present.

Local velocity features of the flow-field are characterized downstream of the runner. The V velocity measured on lines transverse to the draft tube (L1 and L2) display the axial velocity defect due to the vortex structure. Both simulation and experiment show that this velocity defect is most pronounced for the HL case. The velocity defect observed for the BEP case is still present at the center of the passage, but does not show the same strength. The time-averaged V velocity observed for the part load case shows a much broader distribution of due to the large and unsteady nature of the vortex rope. U velocity profiles are much lower in magnitude and show minor levels of discrepancy for the off-design conditions.

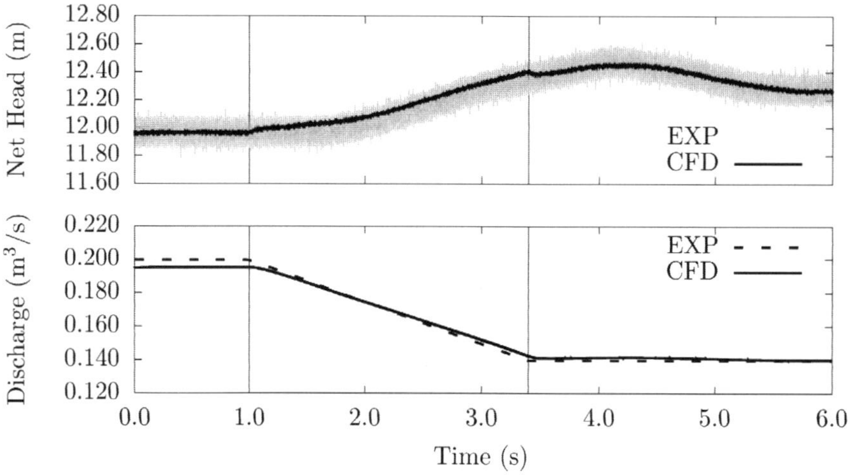

Figure 13: Net head and discharge results of the load reduction operating condition for the experiment and simulation. Vertical lines denote start and end of guide vane rotation.

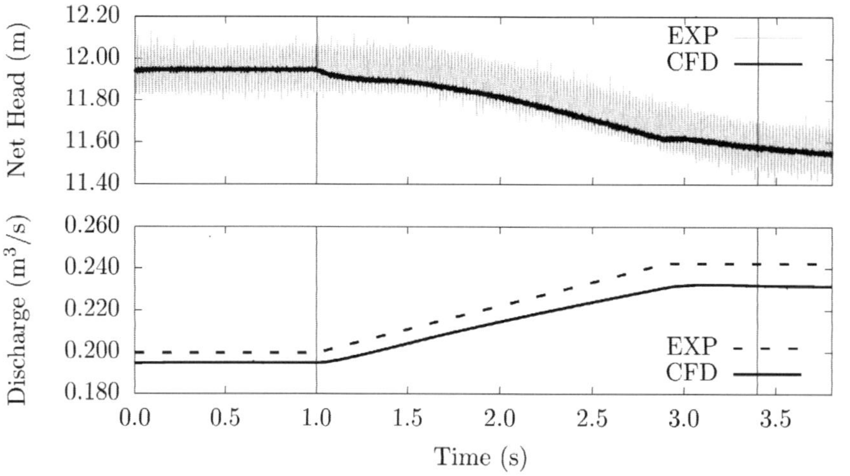

Figure 14: Net head and discharge results of the load acceptance operating condition for the experiment and simulation. Vertical lines denote start and end of guide vane rotation.

Velocity predictions aligned with the runner axis (L3) agree well with experiment for the BEP case. Velocities on L3 show higher levels of discrepancy for the PL and HL cases. A small vortex core is observed for the HL case. Conversely, the vortex structure associated with the PL condition is large, unsteady, and interacts strongly with the draft tube. This complex interaction impacts the relative agreement between simulation and experiment. Streamlines showing velocity magnitude are used to show the relative strength of the vortex rope and it's interaction with the draft tube.

Unsteady simulation from BEP to HL and from BEP to PL was achieved using grid morphing to physically move the guide vanes as a function of time. Both the overall predicted values and unsteady trends agree with experiment. Notably, simulation predicts a sharp change in predicted values at the start and end of guide vane rotation. This is likely due to three factors. First, the fluid was modeled as strictly incompressible, where the physical system will include a small level of compressibility. Second, the simulation domain does not encompass the entire test loop and constant-value boundary conditions are applied. Finally, the function used to specify the vane rotation is non-differentiable, whereas it is likely that the rotation profile imposed in the experiment was smooth.

Once properly validated, simulation is able to accurately predict overall machine performance for both steady and unsteady operation. Additionally, detailed insight into flow-field characteristics offer insight on how turbines may be better designed and where experimental measurements may be taken to be of highest value. Simulation provides an accessible platform for repeatable analysis of machines and can enable design-space exploration, sensitivity analysis, and detailed understanding of in situ operation.

Turbulence models better able to model anisotropic flows such as the vortex rope often outperform more computationally efficient models such as the $k-\omega$ SST model used in this study. A sensitivity to turbulence model with a baseline Large Eddy Simulation (LES) will be performed.

With a turbulence model that is better able to model the complex flow structures of the vortex rope, the long time-scale flow features of the draft tube will be more fully characterized to understand driving factors and how these features impact performance. In support of this effort, it will likely be necessary to extend the computational domain to remove constraints from the system.

References

[1] Renewables 2016 global status report.

[2] Jacek Swiderski, Joseph N Martin, and Richard Norrena. Automated runner blade design optimization process based on cfd verification. *parameters*, 11:n11, 2001.

[3] Jingchun Wu, Katsumasa Shimmei, Kiyohito Tani, Kazuo Niikura, and Joushirou Sato. Cfd-based design optimization for hydro turbines. *Journal of Fluids Engineering*, 129(2):159–168, 2007.

[4] P Drtina and M Sallaberger. Hydraulic turbinesbasic principles and state-of-the-art computational fluid dynamics applications. *Proceedings of the Institution of Mechanical Engineers, Part C: Journal of Mechanical Engineering Science*, 213(1):85–102, 1999.

[5] Sanjay Jain, RP Saini, and Arun Kumar. Cfd approach for prediction of efficiency of francis turbine. In *The 8th International Conference on Hydraulic Efficiency Measurement*, 2010.

[6] Santiago La´n Beatove, Manuel J Garc´a Ruiz, Brian Quintero Arboleda, and Santiago Orrego Bustamante. Cfd numerical simulations of francis turbines. *Revista Facultad de Ingenier´a*, (51):31–40, 2013.

[7] Manoj Kumar Shukla, Rajeev Jain, Vishnu Prasad, and S Shukla. Cfd analysis of 3-d flow for francis turbine. *Int. J. Mech. Eng*, 2:93–100, 2011.

[8] MV Magnoli. Numerical simulation of pressure oscillations in francis turbine runners. *JASS Numerical Simulation of Turbomachinary*, pages 32–50, 2009.

[9] Chirag Trivedi, Michel J Cervantes, BK Gandhi, and Ole G Dahlhaug. Experimental and numerical studies for a high head francis turbine at several operating points. *Journal of Fluids Engineering*, 135(11):111102, 2013.

[10] Kenji Shingai, Nobuaki Okamoto, Yuta Tamura, and Kiyohito Tani. Long-period pressure pulsation estimated in numerical simulations for excessive flow rate condition of francis turbine. *Journal of Fluids Engineering*, 136(7):071105, 2014.

[11] Starccm+® — cfd simulation software. `http://www.cd-adapco.com/products/star-ccm%C2%AE`. Accessed: 2016-09-13.

[12] Norwegian Hydropower Center. Francis-99. `https://www.ntnu.edu/nvks/francis-99`. Accessed: 2016-09-15.

[13] Milovan Peric. Flow simulation using control volumes of arbitrary polyhedral shape. *ERCOFTAC Bulletin*, 62, September 2004.

[14] International Electrotechnical Commission. Hydraulic turbines, storage pumps and pump-turbines: Model acceptance tests. Technical Report IEC 60193, 3, rue de Varemb, PO Box 131, CH-1211 Geneva 20, Switzerland, 1999. Second Edition.

[15] Sir Maurice. Kendall, Alan Stuart, and J. Keith. Ord. *The advanced theory of statistics*. Oxford University Press, 1983.

[16] Colin Heye. Introduction to overset mesh with star-ccm+. In *13th Symposium on Overset Composite Grids and Solution Technology*, October 2016.

Numerical investigation of flow structure and pressure pulsation in the Francis-99 turbine during startup

A Minakov[1,2], **A Sentyabov**[1,2] **and D Platonov**[1,2]

[1] Siberian Federal University, Krasnoyarsk, Svobodniy 79, 660079 Russia
[2] Institute of Thermophysics SB RAS, Novosibirsk, Lavrentyeva 1, 630090 Russia

E-mail: tov-andrey@yandex.ru, sentyabov_a_v@mail.ru, platonov-08@yandex.ru

Abstract. We performed numerical simulation of flow in a laboratory model of a Francis hydroturbine at startup regimes. Numerical technique for calculating of low frequency pressure pulsations in a water turbine is based on the use of DES (k–ω Shear Stress Transport) turbulence model and the approach of "frozen rotor". The structure of the flow behind the runner of turbine was analysed. Shows the effect of flow structure on the frequency and intensity of non-stationary processes in the flow path. Two version of the inlet boundary conditions were considered. The first one corresponded measured time dependence of the discharge. Comparison of the calculation results with the experimental data shows the considerable delay of the discharge in this calculation. Second version corresponded linear approximation of time dependence of the discharge. This calculation shows good agreement with experimental results.

Keywords: Francis turbine, numerical simulation, pressure pulsation, the precession of the vortex rope, turbulence, CFD, LES, RANS and DES models.

1. Introduction

One of the main role of hydroelectric power plants is providing peaking power production. Therefore, quick varying of the operating mode, as startup, shutdown or adjustment of power output, are often required for the hydraulic turbines. In phase of transient between different modes of operation, pulsation phenomena are more intensive than in steady operation [1]. Transient operation results in high amplitude of the dynamic forces on the runner. High pressure pulsations occurs in the part load zone with vortex rope.

Flows in hydroturbine systems are characterized by several features which pose a challenge to the RANS models. The flows in the guide vanes and the rotating runner are dominated by the bounding wall and a strong favorable pressure gradients. The phenomena in a draft tube are different. Here the flow is governed by a strong swirl and a bluff-body recirculation zone behind the runner hub, which is usually reinforced by the negative pressure due to the swirl. The 90 degree bend encountered in elbow draft tubes, with a subsequent circular-to-rectangular change of the cross-section further modify the flow especially at part loads creating very complex vortical patterns with additional recirculation and secondary flows. On the other hand, the walls are smooth, and despite a strong elbow curvature, the wall-adjacent azimuthal fluid velocity and turbulence seem to follow reasonably well the common wall scaling so that computations seem manageable by the common wall-integration schemes or even with using the standard wall-function approach.

Since transients between different modes of turbine operation are very important for the hydraulic unit and accompanied by a large number of complex hydrodynamic phenomena, it is important to be able to simulate dynamic of these processes together. Accordingly, the aim of this paper is to test the numerical algorithm, suitable for simulation of turbine startup process.

Unsteady processes in hydraulic turbines were studied in many papers. The paper [2] presents a mathematical model of hydro power units for different operating conditions, based on the basic version of the software TOPSYS. The simulation and on-site measurements are compared for start-up, no-load operation, normal operation, and load rejection in different control modes (frequency, opening, and power feedback). As a result, the model application is proved trustworthy for simulating different physical quantities of the unit (e.g., guide vane opening, active power, rotation speed, and pressures at volute and draft tube).

The examples of 1D- 3D modeling of transient processes in turbine can be found in works [3 – 5]. In the paper [3] a new approach for numerical simulation of 3D hydraulic turbine flows in transient operating regimes is presented. The method is based on a coupled solution of incompressible RANS equations, runner rotation equation, and water hammer equations. The issue of setting appropriate boundary conditions is considered in detail. As an illustration, the simulation results for runaway process are presented. The evolution of vortex structure and its effect on computed runaway traces are analyzed.

The study Yanna Liu et. al. [4] establishes a mathematical model of the transient process in hydropower stations and presents a new method to calculate the hydraulic turbine boundary based on an error function of the rotational speed. The mathematical derivation shows that the error function along the equal-opening characteristic curve is monotonic and has opposite signs at the two sides, which means that a unique solution exists to make the error function null.

The paper [5] presents the simulation and the analysis of the transient process of a Francis turbine during the load rejection by employing a one-dimensional and three-dimensional (1-D-3-D) coupling approach. The coupling is realized by partly overlapping the 1-D and 3-D parts, the water hammer wave is modelled by defining the pressure dependent density, and the guide vane closure is treated by a dynamic mesh method. To verify the results of the coupling approach, the transient parameters for both typical models and a real power station are compared with the data obtained by the 1-D approach, and good agreements are found. To investigate the differences between the transient and steady states at the corresponding operating parameters, the flow characteristics inside a turbine of the real power station are simulated by both transient and steady methods, and the results are analysed in details.

2. Mathematical model and numerical method

The computations here reported were performed with the ANSYS-FLUENT code using the DDES method (based on the *k*-ω SST Menter's model). The basic equations of the mathematical models express the conservation laws in the rotating reference frame.
The continuity equation (conservation of mass):

$$\frac{\partial u_i}{\partial x_i} = 0 \tag{1}$$

Momentum equations (conservation of momentum) in a rotating reference frame for absolute velocities:

$$\frac{\partial \rho u_i}{\partial t} + \frac{\partial}{\partial x_j}\left(\rho u_j^r u_i\right) = -\frac{\partial p}{\partial x_i} + \frac{\partial}{\partial x_j}(\tau_{ij}^{m} + \tau_{ij}^{t}) - \rho \varepsilon_{ijk} \Omega_j u_k \tag{2}$$

Where: u_i – absolute velocity components, u_j^r – relative velocity components, τ_{ij}^{m} – viscous stress tensor, τ_{ij}^{t} – turbulent stress tensor, Ω – angular velocity of runner rotation, p – static pressure, ρ – density, ε_{ijk} – Levi-Civita symbol.

Components of the viscous stress tensor are defined as:

$$\tau_{ij}^{\mathrm{m}} = \mu\left(\frac{\partial u_i}{\partial x_j} + \frac{\partial u_j}{\partial x_i}\right) \tag{3}$$

where μ is dynamic molecular viscosity. In constructing two-equation models of turbulence for defining the components of Reynolds' stress tensor τ_{ij}^{t}, Boussinesq's hypothesis of isotropic turbulent viscosity is used:

$$\tau_{ij}^{\mathrm{t}} = \mu_{\mathrm{t}}\left[\left(\frac{\partial u_i}{\partial x_j} + \frac{\partial u_j}{\partial x_i}\right) - \frac{2}{3}\delta_{ij}\rho k\right] \tag{4}$$

In this paper we use the DES approach based on k-ω SST model [6]. The dissipation term of the turbulent kinetic energy is modified for the DES turbulence model as described in [7]. In this model dissipative term in k-equation is modified by means of switcher F_{DES}:

$$\frac{\partial(\rho k)}{\partial t} + \frac{\partial}{\partial x_j}\left(\rho u_j k\right) = \frac{\partial}{\partial x_j}\left(\Gamma_k \frac{\partial k}{\partial x_j}\right) + G_k - Y_k \cdot F_{\mathrm{DES}}, \tag{5}$$

$$F_{\mathrm{DES}} = \max\left(\frac{l_{\mathrm{t}}}{C_{\mathrm{DES}}\Delta}, 1\right), \qquad l_{\mathrm{t}} = \frac{k^{1/2}}{\beta^* \omega}, \qquad C_{\mathrm{DES}} = 0.61$$

where: l_{t} – turbulent length scale, C_{DES} – empirical constant, and Δ is defined as the maximum of three sizes of the control volume $\Delta_x, \Delta_y, \Delta_z$.

A specific issue in the computer simulation of hydraulic turbines is the treatment of the rotating impeller and the rotor-stator interaction. Several approaches can be found in the literature, i.e. dynamic, sliding and moving grid methods, and those based on a moving reference frame. The latter is the most common and the simplest way to model the runner rotation. It assumes that the runner is fixed and the equations are solved in a rotating reference frame. This formulation is often referred to as the "frozen rotor" approach. In this paper, the modelling of the runner rotation was performed in the rotated reference frame for the runner zone. The earlier test calculations proved that this approach is credible for describing the integral flow characteristics including the dominant flow pulsations [8 – 10]. The "frozen rotor" (known also as "moving reference frame") is a well-established and widely used method for treating stator-rotor interaction. In this approach the rotating and the stationary parts have fixed relative positions. A frame transformation is done to include the rotating effect on the rotating sections. Simultaneous computations of stator and rotor flow using sliding meshes is undoubtedly superior and more exact, but it is computationally more demanding, it requires longer iterations, and it is less robust. The frozen rotor approach is more practical, economical and robust; our earlier test for a similar configuration showed a very small, almost marginal difference between the results obtained by the two approaches.

The computations were performed using the finite-volume method on unstructured polyhedral grids. The coupling of the velocity and pressure fields for incompressible flow was ensured using the SIMPLE-C procedure. The second-order central difference scheme was used for DES. The time derivatives were approximated by an implicit second-order scheme. The time step was set from the condition of averaged CFL < 2.

3. Computational domain, boundary conditions and grids

The computational domain included the runner and the draft tube (see Fig.1). The 3D geometry of the turbine has been taken on the website workshop Francis (2016). The computations have been carried out using unstructured grids with the total number of 7.06 million nodes for the whole domain (Fig.2). The distance of the wall-nearest grid nodes y^+, was approximately from 40 to 1000. Despite the large number of nodes, this mesh is insufficient for the correct resolution of the boundary layers. Nevertheless, the mesh is suitable for simulation of the large-scale vortices in the draft tube [6].

The inlet boundary conditions were set at the runner inlet by means of time-dependent functions of the radial and tangential velocities (Fig. 3, version 1). The radial velocity component was obtained from the approximated time dependence of the experimental value of discharge (Fig. 4, version 1), and the tangential velocity component was obtained from the radial velocity and experimental value of the guide vanes angle (Fig. 5). Experimental discharge was approximated by means of polynomial (Fig. 4, version 1). However, for the transient measurements, there is some delay in the flowmeter. For this reason, another approximation of the discharge evolution was also tested. Second version of the discharge time dependence was is linear between 2 and 10 s (Fig 4). Such period correspond to the experimental velocity data.

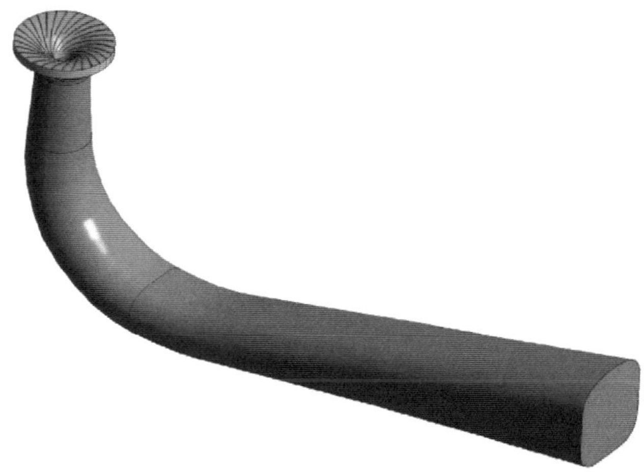

Figure. 1. The computational domain.

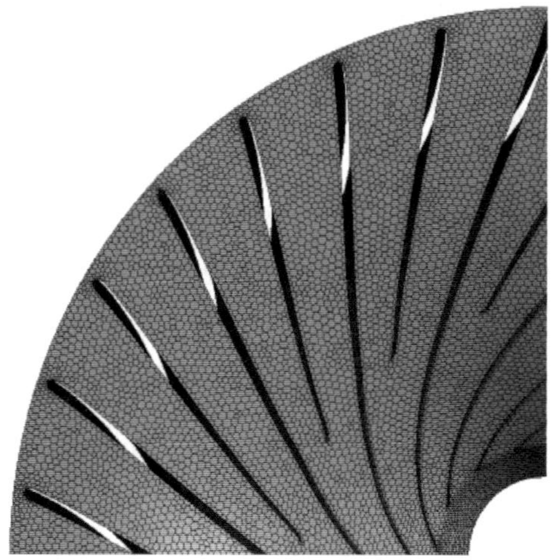

Figure. 2. Views of the computational grid with 7 million nodes in the turbine mode (the draft tube and the runner).

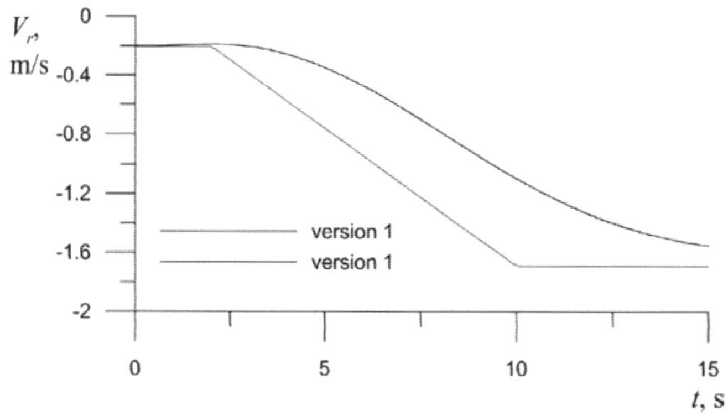

a)

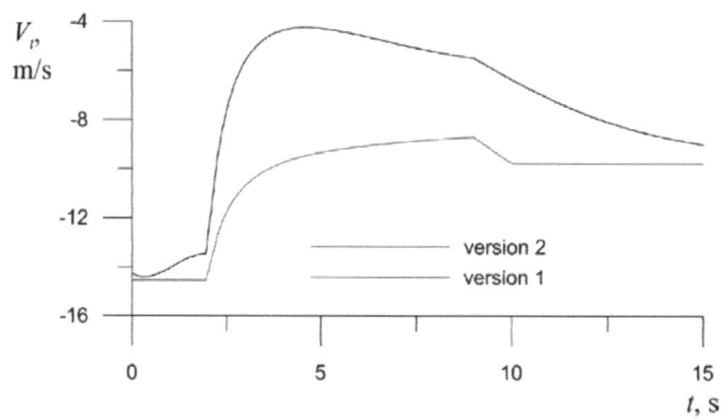

b)

Figure. 3. Velocity components on the inlet: a) radial velocity, b) tangential velocity. Version 1 is based on measured discharge, version 2 is based on linear approximation of the discharge

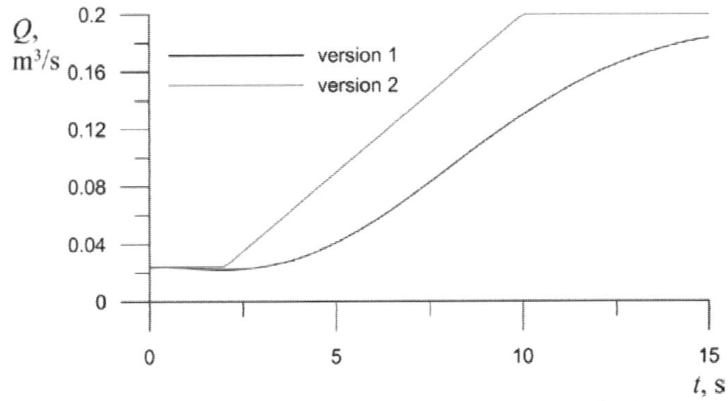

Figure. 4. Approximated time dependence of the discharge.

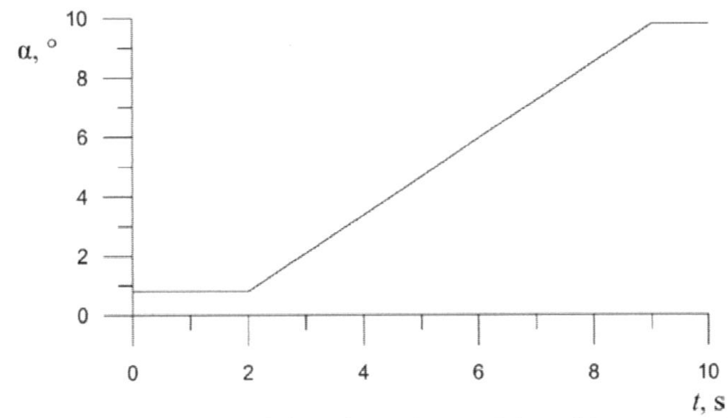

Figure. 5. Approximated time dependence of the guide vanes angle.

4. The results of the startup simulation in the Francis-99 turbine

As shown in Fig. 6, at the initial time at the small guide wicket gate there is opening long vortex rope under the runner and in the draft tube elbow (Fig. 6, time marks correspond Fig. 3 – 5). Calculated pressure pulsations closely agree with the experimental data (Fig. 7, 8). The vortex rope precession induces intensive low-frequency pressure pulsations in the draft tube at $f \approx 1.5$ Hz which are clearly seen for the initial stage of the startup (Fig. 9). Due to the long vortex rope, that reaches draft tube elbow, pressure pulsations are very intensive at this stage. Likely, interaction of the vortex rope and the draft tube elbow induces synchronous pressure pulsation in addition to asynchronous pulsations generated by vortex rope precession [10]. At the end of the startup the experimental spectrum contains a number of high frequencies (Fig. 9).

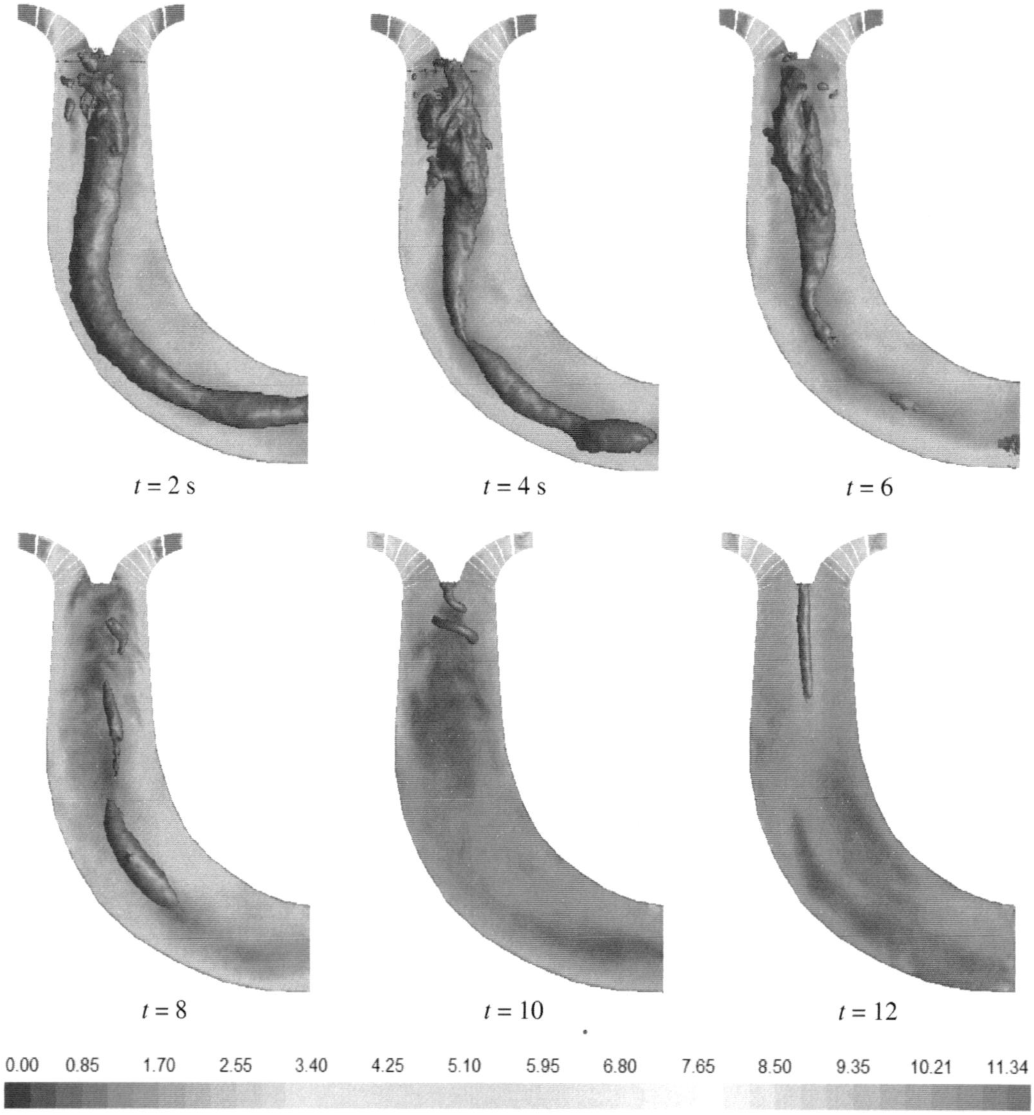

Figure. 6. Vortex evolution in the draft tube visualized by pressure iso-surface (red), velocity magnitude in the central longitudinal cross-section (obtained for the calculation No 1).

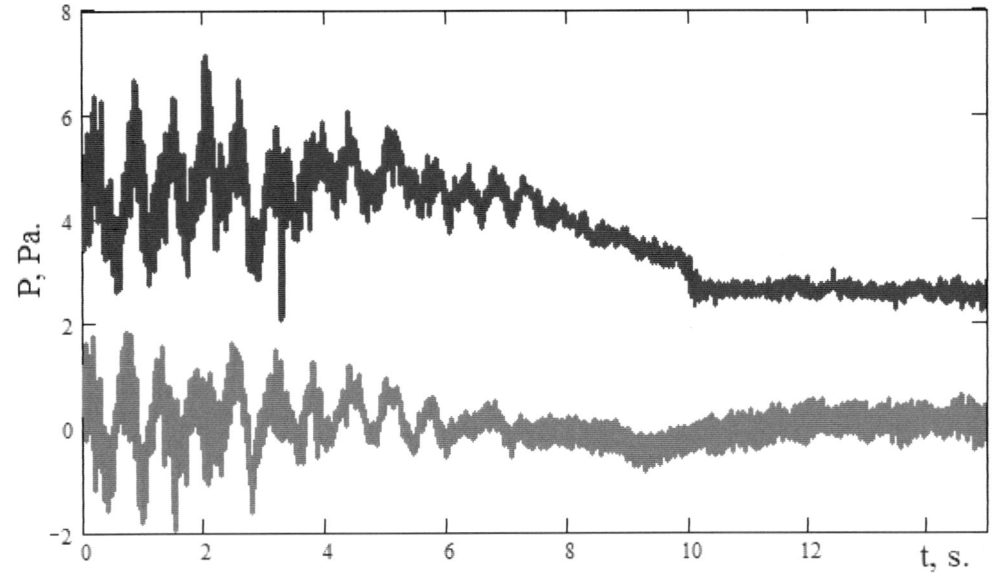

Figure. 7. Pressure pulsations at the point DT5 (blue line – simulation, red line - experiment). Calculation No 2

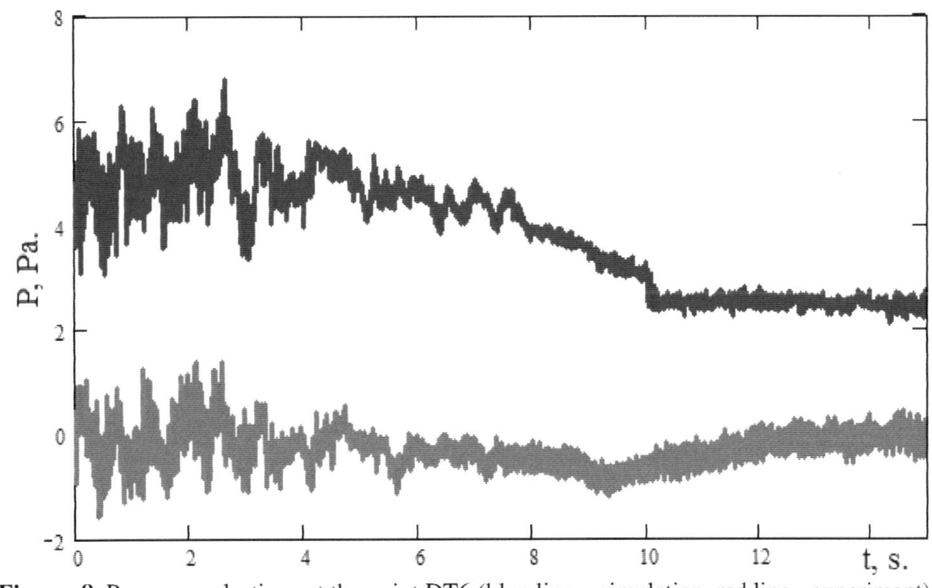

Figure. 8. Pressure pulsations at the point DT6 (blue line – simulation, red line - experiment). Calculation No 2

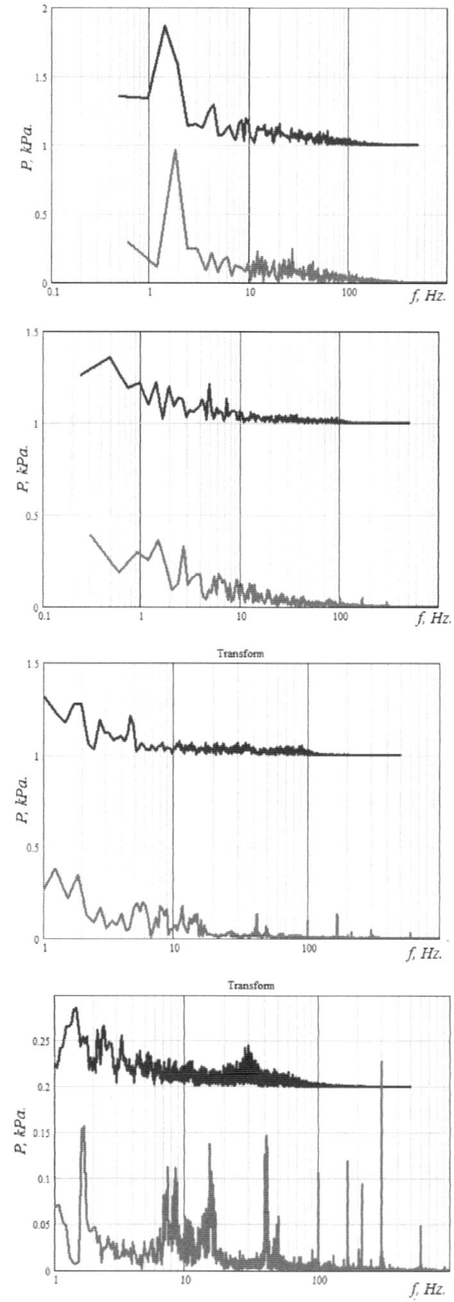

a) b) c) d)

Figure. 9. Spectra of the pressure pulsations at the point DT5 (blue line – simulation, red line - experiment): a) time window is 0-2 s; b) time window is 2-4 s; c) time window is 4-8 s; d) time window is 8-24 s. Calculation No 2

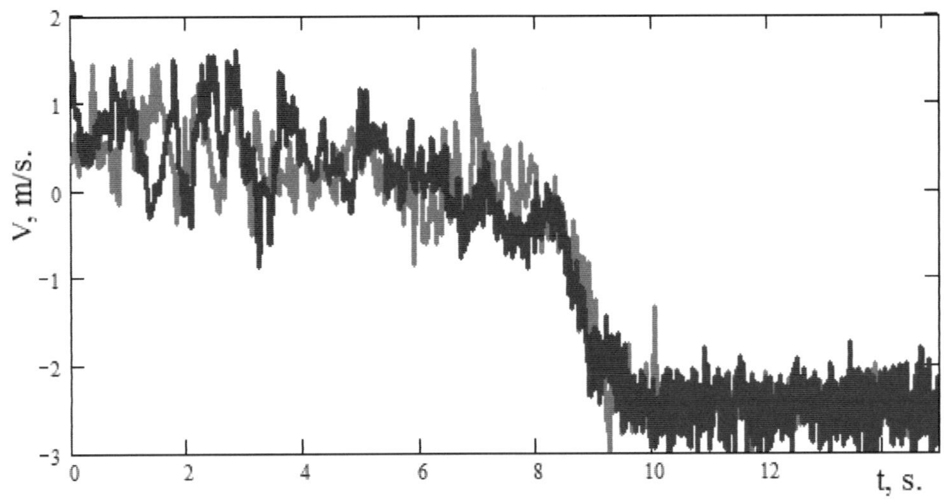

Figure. 10. Time dependence of the axial velocity near the axis of the draft tube, point #215: x = -0.75 mm, y = -3.87, z = -458.60 mm (blue line – simulation, red line – experiment). Calculation No 2

As wicket gate opening increase and, accordingly the flow swirl number decreases, intensity of the vortex rope reduces (Fig. 6, t = 4 s). As a result, magnitude of the pressure pulsations decreases. Then (Fig. 6, 8 – 10 s), intensive vortex rope in the draft tube collapses and recirculation zone under the runner vanishes (Fig. 11, 8 – 10 s). At this stage, only high-frequency pressure pulsations, associated with the runner rotation, remain in the experimental pressure pulsations but these pulsations were not considered in the calculation (Fig. 9d). Near the best efficiency point, the vortex under the runner has straight form (Fig. 6, t = 12 s) and intensity of the pressure pulsations reduces to the minimum value (Fig. 7, 8).

Comparison of the velocity profiles in the draft tube are shown in Fig. 11. The figure presents calculated and experimental velocity profiles averaged during 2 s. Such period is rather long to include many computational and experimental time steps and is rather short compared to time of the startup. Satisfactory agreement of the calculated and experimental results can be seen in the figure. As Fig. 11 shows, before the turbine startup and at initial transient stage (t = 0 – 2 s, t = 4 – 6 s) there is a wide zone of the backward flow in the draft tube. This recirculation zone forms due to high swirl number of the flow under the runner and, accordingly, precessed vortex rope in the draft tube. At the end of the startup transient process, calculation No 1 shows considerable discrepancy with the experimental data. The plot reveals that the calculated discharge is smaller than the experimental one. It is due to delay of the flowmeter. As Fig. 11 shows, calculation No 2 describes the startup well.

IOP Conf. Series: Journal of Physics: Conf. Series **782**(2017) 012004 doi:10.1088/1742-6596/782/1/01200

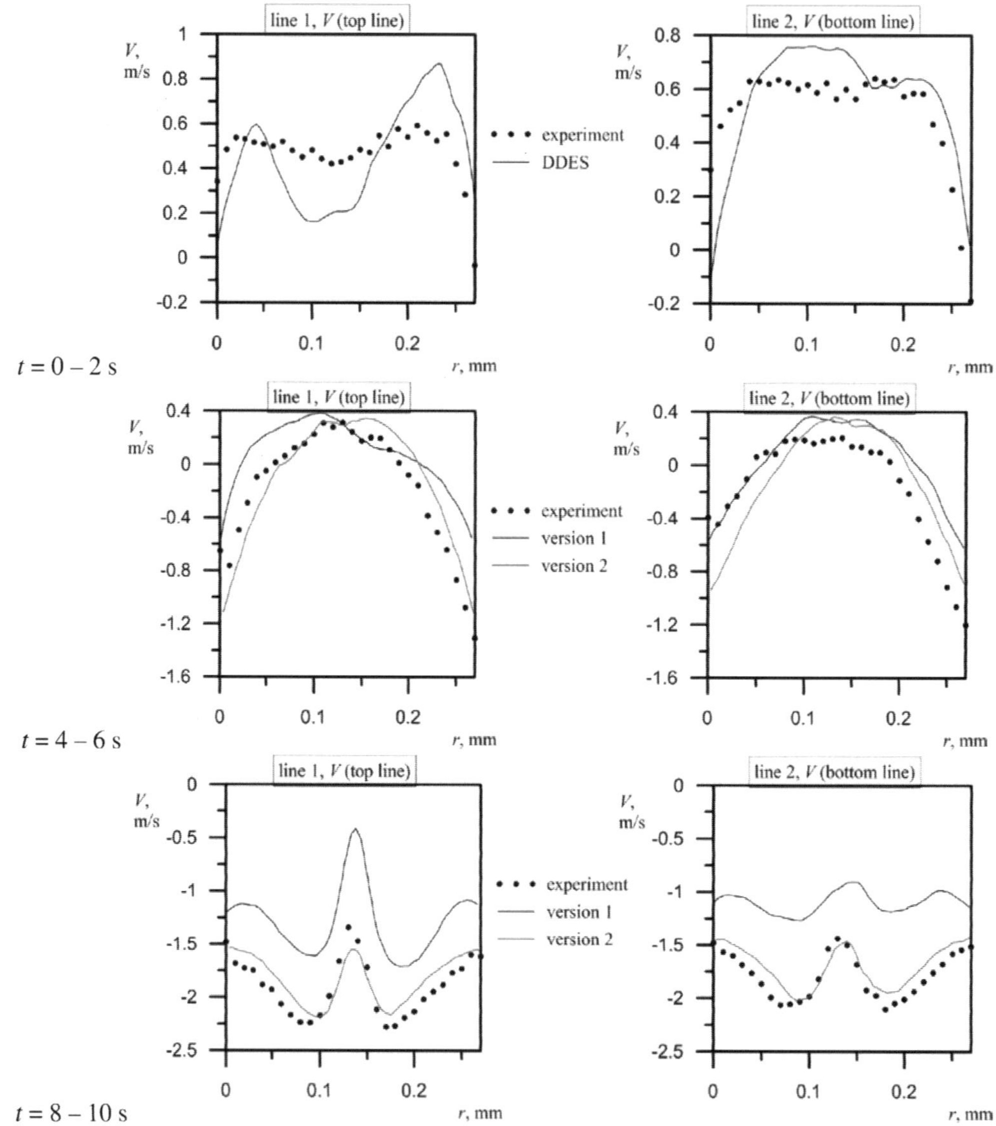

Figure. 11. Averaged axial velocity component in the draft tube.

6. Conclusions

Thus, as the opening of wicket gate increase, the flow under the runner pass different stages corresponding to the stable-state operation. At low wicket gate opening, the flow in the draft tube is highly swirled, wide recirculation zone forms under the runner and vortex rope rotates around it. The vortex rope has several period of the rotation during the startup and induce intensive low-frequency pressure pulsations. Then, the vortex rope gradually destroys and pressure pulsations becomes non-periodical. At last, only weak straight vortex core remains under the runner and the pressure pulsations reaches minimal magnitude.

Good agreement between the experimental and calculated data shows that the algorithm is suitable for the simulating of the transient processes in hydraulic turbines. Since the main influence on the unsteady flow processes has the flow in the draft tube, the flow in the guide vanes and the spiral case can in neglected in the simplest approach, and only the runner and the draft tube can be considered. In addition, detached eddy simulation is suitable for the correct resolving of the dominant vortices under the runner. Comparison of the calculated and experimental unsteady behaviors of the velocity in the monitoring points in the draft tube has shown that the calculation is in very good agreement with experiment both in amplitude and frequency fluctuations of velocity during start-up of the turbine.

In addition it is necessary to note one more interesting feature. In this paper, modeling of the opening of the guide device in the startup process is not carried out. Firstly, as the inlet boundary conditions time dependences of the radial and tangential velocity components, obtained from the experimental values of the opening angle of the guide vanes and flow discharge, were used. Results of the simulation in this formulation showed that the calculated profiles of axial velocity is significantly different from the experimental data. Therefore, the discharge time dependence has been adjusted. It was used the assumption that the discharge linearly increases from a minimum to a maximum value during startup period of time 8s. This implicitly agrees with the delay in measured discharge. Also it is confirmed by experimental measurements of the velocities in the monitoring points depending on time. Calculations showed that second version of the discharge dependence provides a very good agreement of the velocity profiles during the startup of the turbine.

Acknowledgements
This work was supported by Russian Foundation for Basic Research 16-48-240707.

References

[1] Doerfler P, Sick M and Coutu A 2013 Flow-Induced Pulsation and Vibration in Hydroelectric Machinery. *Engineers Guidebook for Planning, Design and Troubleshooting*. London: Springer

[2] Weijia Yang, Jiandong Yang, Wencheng Guo, Wei Zeng, Chao Wang, Linn Saarinen and Per Norrlund 2015 A Mathematical Model and Its Application for Hydro Power Units under Different Operating Conditions *Energies* **8** 10260-10275; doi:10.3390/en80910260.

[3] Cherny S, Chirkov D, Bannikov D, Lapin V, Skorospelov V, Eshkunova I and Avdushenko A 2010 3D numerical simulation of transient processes in hydraulic turbines. *IOP Conf. Series: Earth and Environmental Science* **12** 012071 doi:10.1088/1755-1315/12/1/012071

[4] Yanna Liu, Jiandong Yang, Jiebin Yang, Chao Wang and Wei Zeng 2016 Transient Simulations in Hydropower Stations Based on a Novel Turbine Boundary. *Mathematical Problems in Engineering* **Volume 2016**, Article ID 1504659, 13 pages. http://dx.doi.org/10.1155/2016/1504659

ZHANG Xiao-xi, CHENG Yong-Guang, YANG Jian-Dong, XIA Lin-Sheng, LAI Xu.

[5] Simulation of the load rejection transient process of a Francis turbine by using a 1-D-3-D coupling approach. Journal of Hydrodynamics 26(5),·2014,26(5):715-724 DOI: 10.1016/S1001-6058(14)60080-9

[6] Menter F R 1994 Two Equation Eddy Viscosity Turbulence Models for Engineering Applications *AIAA J.* **32**, № 8, 1598-1605

[7] Strelets M 2001 Detached Eddy Simulation of Massively Separated Flows. *AIIA Paper.* 2001-0879

[8] Minakov A V, Platonov D V, Dekterev A A, Sentyabov A V and Zakharov A V 2015 The numerical simulation of low frequency pressure pulsations in the high-head Francis turbine *Computers and Fluids* **111** pp. 197-205, DOI:10.1016/j.compfluid.2015.01.007

[9] Minakov A V, Platonov D V, Dekterev A A, Sentyabov A V and Zakharov A V 2015 The analysis of unsteady flow structure and low frequency pressure pulsations in the high-head Francis turbines *International Journal of Heat and Fluid Flow* **53** pp. 183-194, DOI: 10.1016/j.ijheatfluidflow.2015.04.001

[10] Gavrilov A A, Sentyabov A V, Dekterev A A and Hanjalić K 2016 Vortical structures and pressure pulsations in draft tube of a Francis-99 turbine at part load: RANS and hybrid RANS/LES analysis *International Journal of Heat and Fluid Flow* **(June)**

doi:10.1016/j.ijheatfluidflow.2016.05.007
[11] Sentyabov A V, Gavrilov A A, Dekterev A A and Minakov A V 2014 Numerical investigation of the vortex core precession in a model hydro turbine with the aid of hybrid methods for computation of turbulent flows *Thermophysics and Aeromechanics* **21(6)** pp. 707 – 717.

Francis-99 turbine numerical flow simulation of steady state operation using RANS and RANS/LES turbulence model

A Minakov[1,2], D Platonov[1,2], A Sentyabov[1,2] and A Gavrilov[1,2]

[1] Siberian Federal University, Krasnoyarsk, Svobodniy 79, 660079 Russia
[2] Institute of Thermophysics SB RAS, Novosibirsk, Lavrentyeva 1, 630090 Russia

E-mail: tov-andrey@yandex.ru, platonov-08@yandex.ru, sentyabov_a_v@mail.ru, gavand@yandex.ru

Abstract. We performed numerical simulation of flow in a laboratory model of a Francis hydroturbine at three regimes, using two eddy-viscosity- (EVM) and a Reynolds stress (RSM) RANS models (realizable k-ε, k-ω SST, LRR) and detached-eddy-simulations (DES), as well as large-eddy simulations (LES). Comparison of calculation results with the experimental data was carried out. Unlike the linear EVMs, the RSM, DES, and LES reproduced well the mean velocity components, and pressure pulsations in the diffusor draft tube. Despite relatively coarse meshes and insufficient resolution of the near-wall region, LES, DES also reproduced well the intrinsic flow unsteadiness and the dominant flow structures and the associated pressure pulsations in the draft tube.

Keywords: Francis turbine, numerical simulation, pressure pulsation, the precession of the vortex rope, turbulence, CFD, LES, RANS and DES models.

1. Introduction

An important task of hydropower plant is the regulation of power in the energy system. During load changing, hydraulic units repeatedly undergo off-design modes of operation. In these modes, the flow remains essentially swirling after passing through the water turbine runner. The instability of swirling flow results in the appearance of intense low-frequency hydrodynamic pulsations, which threaten the safety and reliability of turbine structure.

The improvement of hydraulic machines efficiency and failure tolerance is impossible without studying the physical mechanisms of hydrodynamic processes, among which an important role is played by the transient phenomena associated with the formation of large-scale vortex structures. One of the mechanisms responsible for generating flow pulsations is the precessing vortex rope. It is formed downstream the runner under part-load or over-load conditions when the flow has a large residual swirl after passing through the runner [1–5].

The precession of vortex rope can be a serious danger for hydraulic equipment due to powerful flow pulsations that lead to strong vibrations of turbine structure. It the case of resonance, it can result in the turbine structural failure. Pressure pulsations generated by precessing vortex rope may also affect the cavitation process and enhance cavitation erosion. To predict the resonance phenomena and find the ways of suppressing instability, one needs the detailed information about the characteristics of pulsation regimes and flow structure. It should be noted that the working approaches must meet the requirement of minimizing energy losses (increasing the efficiency of water turbine). It can be realized only through in-depth understanding of hydrodynamic processes occurring in the flow path of a water turbine.

The precession of vortex rope has been studied for quite a long time. These issues have long been in focus of the experimental research complemented by some simplified analytics, but a full account of the three-dimensional time dynamics has remained to a large extent beyond the reach of even most advanced laser-based measurements and diagnostics techniques. Computer simulations have been seen as a potential tool for capturing the subtleties of the development of instabilities and associated vortical and turbulence structures. Such information can be obtained by direct and large-eddy simulations (DNS, LES) over a range of important scales. However, the accuracy of DNS and LES depends much on the numerical resolution. For complex configurations the Reynolds-averaged Navier-Stokes (RANS) approach, being computationally less demanding, has been considered as a more rational option for industrial purposes. However, because if its empirical contents, the RANS have not been regarded as a trustful research tool, and moreover, their credibility depends much on the choice of the turbulence model to close the time- or ensemble-averaged conservation equations.

Despite the extensive literature (e.g. Doerfler [2], Alligne et al. [6], comprehensive experimental data for hydroturbines are still scarce, apart from mean velocity and second- moment turbulence statistics measured at selected locations in laboratory models of some popular turbine types [7-8]. On the other hand, there are still many open questions regarding the numerical simulation of flows in turbines of realistic configuration and practical relevance. Because of typically very high Re numbers, DNS and even LES, requiring enormous numerical mesh and computer resources, are usually beyond the reach of designers, but also of many researchers, leaving the URANS or hybrid RANS/LES as the only viable routes. Here the first challenge is related to selecting the appropriate RANS model that can reproduce satisfactorily the unsteady flow and the dynamics of at least the dominant large-scale vortex structures subjected to strong swirls, curvature and adverse pressure gradients in complex geometric shapes. The popular linear eddy-viscosity modes (k-ε, k-ω, and their many "improved" variants) lack on sufficient physics and have shown to perform poorly in describing such flows.

We report here the results of simulation of flow in a Francis-99 hydraulic turbine at three regimes using several RANS models, a DES (detached-eddy simulations) and LES. The models were benchmarked against the experimental data. We discuss the flow patterns and vortical structure in the turbine draft tube and its impact on flow stability and pressure pulsations, especially at part loads characterized by unsteady helix ropes.

2. Turbulence models

Flows in hydroturbine systems are characterized by several features which pose a challenge to the RANS models. The flows in the guide vanes and the rotating impeller are dominated by the bounding wall – blades and housing, and a strong favourable pressure gradient which for the design operating conditions (no flow separation) can be handled reasonably well with a well-tuned wall-integration (WIN) model that (for the runner) accounts properly for the rotation effects. Accurate predictions of the blade-tip leakage, the follow-up vortical wakes and flow separation on blades at part loads, require more advanced, anisotropy-accounting RANS models. The phenomena in a draft tube are different. Here the flow is governed by a strong swirl and a bluff-body recirculation zone behind the impeller hub, which is usually reinforced by the negative pressure due to the swirl. The 90 degree bend encountered in elbow draft tubes, with a subsequent circular-to-rectangular change of the cross-section further modify the flow especially at part loads creating very complex vortical patterns with additional recirculation and secondary flows. On the other hand, the walls are smooth, and despite a strong elbow curvature, the wall-adjacent azimuthal fluid velocity and turbulence seem to follow reasonably well the common wall scaling so that computations seem manageable by the common wall-integration schemes or, according to some publications, even with using the standard wall-function approach. The apparent success of relatively standard wall treatment could in part be explained by strong radial pressure gradients and by inviscid interactions that diminish the wall effects as well as the role of the employed turbulence model on the dominant vortical structures and the consequent mean flow pattern.

The above brief overview of some major phenomena encountered in hydromachinery illustrate the importance of choosing the proper RANS model, which depends on which phenomena are in the focus of the computations. As stated above, our focus is on the draft tube and, specifically, on the strong and coherent swirling structures, primarily the precessing vortex cores and especially on the undesirable twin-ropes and the associated strong pressure pulsations at part loads. The linear eddy-viscosity (LEVM) turbulence models, widely spread in engineering calculations, are known to describe poorly most features listed above. Nevertheless, we tested some popular models, all validated against a well-resolved dynamic LES, aimed at scrutinizing the ability of various models to capture the above mentioned specific structures and pressure oscillation in the considered draft-tube model.

The computations here reported were performed with the ANSYS-FLUENT code using the built-in two popular eddy-viscosity (k-ω SST and the realizable k-ε), a Re-stress model (Launder-Reece and Rodi, LRR, with the linear pressure strain and wall-echo terms), and a DES (based on the k–ω SST Menter's model). In parallel, for reference, LES with the WALE subgrid-scale viscosity model was carried out on finer grids.

The high-Re-number realizable k-ε model and the basic RSM were solved using the standard wall functions. A test with the enhanced wall treatment and the non-equilibrium wall function available in the FLUENT code produced no significant differences in the results. For the k–ω SST we used the enhanced-wall-treatment ω-equation model (EWT-ω).

A specific issue in the computer simulation of hydraulic turbines is the treatment of the rotating impeller and the rotor-stator interaction. Several approaches can be found in the literature, i.e. dynamic, sliding and moving grid methods, and those based on a moving reference frame. The latter is the most common and the simplest way to model the runner rotation. It assumes that the runner is fixed and the equations are solved in a rotating reference frame. This formulation is often referred to as the "frozen rotor" approach. In this paper, the modelling of the runner rotation was performed in the rotated reference frame for the runner zone. The obtained results are then rotated with the runner rotation speed and as such imposed as the inflow field as the inlet into the draft tube. The earlier test calculations proved that this approach is credible for describing the integral flow characteristics including the dominant flow pulsations [4, 5, 8 – 10]. A comparison with the computationally more demanding method using sliding meshes showed that for this type flows with a focus on draft tube dynamics the results are almost the same.

The computations were performed using the finite-volume method on unstructured grids. The coupling of the velocity and pressure fields for incompressible flow was ensured using the SIMPLE-C procedure. For the URANS models the convective terms in all equation were discretized by a second-order upwind scheme, whereas the second-order central difference scheme was used for LES and DES (in the LES region). The time derivatives were approximated by an implicit second-order scheme. The time step was set from the condition CFL<2, but in the bulk of the draft tube the CFL number was less than 1.0. Some peaks exceeding the value of 2.0 appeared close to the guide-vane and impeller blades. The statistics was gathered for each run during about 10 sec of real time.

3. Computational domain, boundary conditions and grids

The computational domain included all the elements constituting the turbine system (spiral casing, guide vane, impeller and draft tube) see (Fig.1). The 3D geometry of the turbine has been taken on the website workshop Francis (2016). Since the intake channel was not considered in the calculations, the boundary conditions were set at the inlet to the turbine spiral casing by imposing a total head measured in the experiment and taken from Table 1. As the LES method requires initial time-dependent forcing mimicking turbulence, random velocity fluctuations were imposed at the inlet, generated by the method proposed in [11].

IOP Conf. Series: Journal of Physics: Conf. Series 782 (2017) 012005 doi:10.1088/1742-6596/782/1/01

Table 1. Acquired flow parameters and other quantities during steady state measurements.

Parameter	PL	BEP	HL
Guide vane angle (°)	6.72	9.84	12.43
Net head (m)	11.87	11.94	11.88
Discharge ($m^3\ s^{-1}$)	0.13962	0.19959	0.24246
Torque to the generator (Nm)	416.39	616.13	740.54
Friction torque (Nm)	4.40	4.52	3.85
Runner angular speed (rpm)	332.84	332.59	332.59
Casing inlet pressure-abs (kPa)	218.08	215.57	212.38
Draft tube outlet pressure-abs (kPa)	113.17	111.13	109.59
Hydraulic efficiency (%)	90.13	92.39	91.71
Water density (kg m^{-3})	999.8	999.8	999.8
Kinematic viscosity ($m^2\ s^{-1}$)	9.57E-7	9.57E-7	9.57E-7
Gravity (m s^{-2})	9.82	9.82	9.82

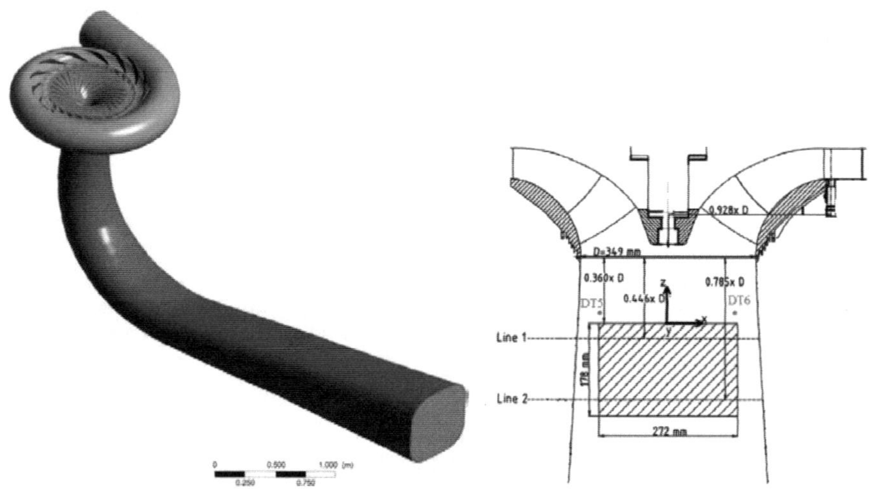

Figure. 1. The computational domain and scheme.

The computations have been carried out using unstructured grids with the total number of 5.12 and to 14.28 mln. nodes for the whole domain (Fig.2). The average distance of the wall-nearest grid nodes normalized by the wall units, y_1^+, for the 5.12 mln. grid was 150, for the 14.28M grid y_1^+ was 89. As seen, both grids are coarse grid for the LES method.

To check further the grid resolution, the characteristic cell size $\Delta = (\Delta x \times \Delta y \times \Delta z)^{1/3}$ was compared with the Kolmogorov dissipative scale $L_K = \left(\nu^3 / \varepsilon\right)^{1/4}$, and the generalized Corrsin "shear" scale

$L_s = \left(\varepsilon / S^3\right)^{1/2}$ (where ν is the kinematic viscosity, ε its dissipation rate (obtained from RSM), and $S = (S_{ij}S_{ij})^{1/2}$ is the strain modulus. The shear scale, characterizing the scale of the turbulence energy production, was found by Picano and Hanjalić [12] to serve well as a criterion for mesh resolution in free flows: when the grid-cell size was kept smaller than the shear scale (nominally $\Delta / L_s < 1.0$, relevant primarily in high-shear regions), the LES of a free round jet reproduced well the mean flow and turbulence statistics at very high Re number using a very coarse grid despite the cell sizes being two order of magnitude larger than the Kolmogorov scale. Fig. 3 shows the distribution of Δ/L_K and Δ/L_s in the central section of the turbine calculated by RSM model for part load (PL) regime for grid 5.12M.

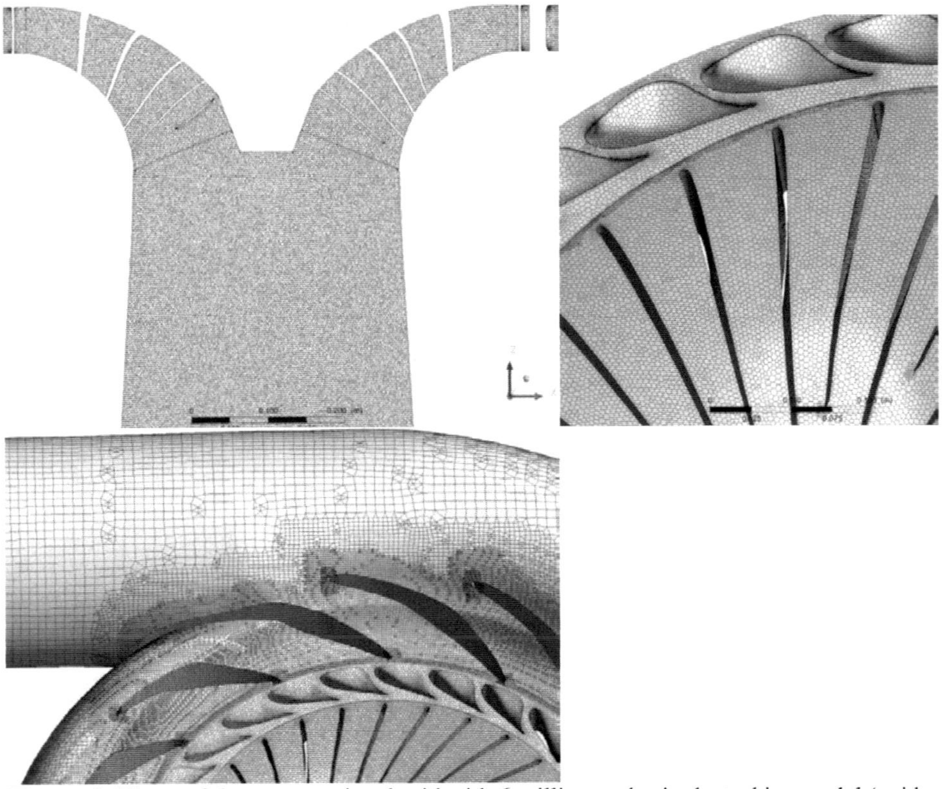

Figure. 2. Views of the computational grid with 6 million nodes in the turbine model (guide vanes and runner) and its draft tube.

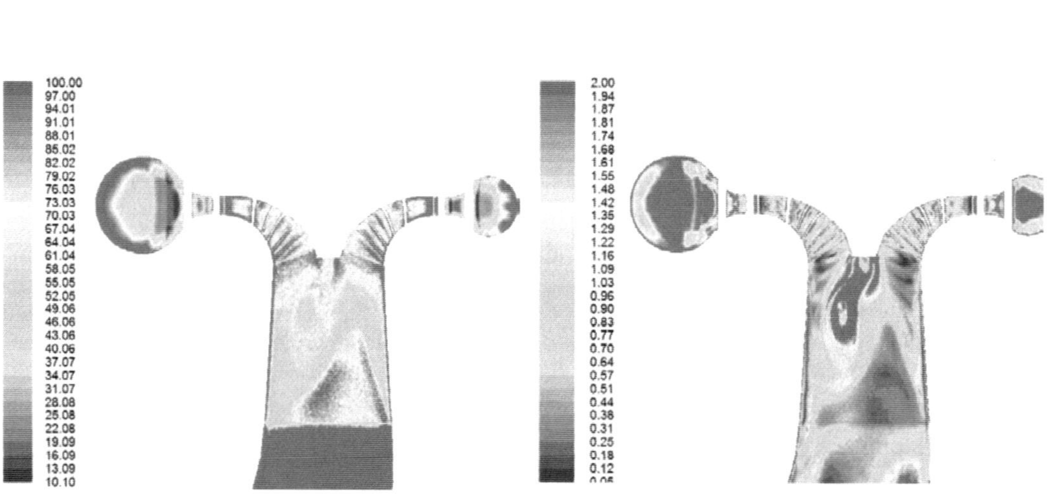

a) *b)*

Figure. 3. Grid size compared with turbulence length scales (a) Δ/L_K and (b) Δ/L_s in the middle plane turbine for 5.12M grid.

As can be seen, the ratio Δ / L_K of around 40 to 80. Interestingly, the cell-size to shear-scale ration Δ / L_S is below 1.0, except in the core region, and of course very close to the wall. This is in agreement with the findings in [12], which shows that the LES method can return satisfactory predictions when Δ / L_S is less than 1.0.

4. The results of simulations flow in the Francis-99 turbine

This section presents the results of simulation of the flow in the turbine Francis-99. Three operation point proposed for the Francis-99 (2016) workshop were calculated: part load (PL), best efficiency point (BEP) and high load (HL). The Fig. 4 shows axial velocity in central cross-section of the turbine for three operation points. Flow at HL and BEP modes are similar: a narrow concentrated vortex is formed under the runner (see Fig. 5). At PL regime axial flow is close to the draft tube wall. At this operation point a wide recirculation zone is formed at the axis in the draft tube cone. It is indicate that a strong vortex rope precesses under partial load, since at under such conditions the flow has a considerable residual swirl. Vortex breakdown in such a flow leads to the formation of the recirculation zone at the flow axis and the rotation of the corkscrew vortex rope around it. Fig. 5 shows selected snapshots of the typical vortex ropes and their structures behind the runner, captured by DES model. The vortex ropes are identified by the instantaneous iso-surfaces of pressure.

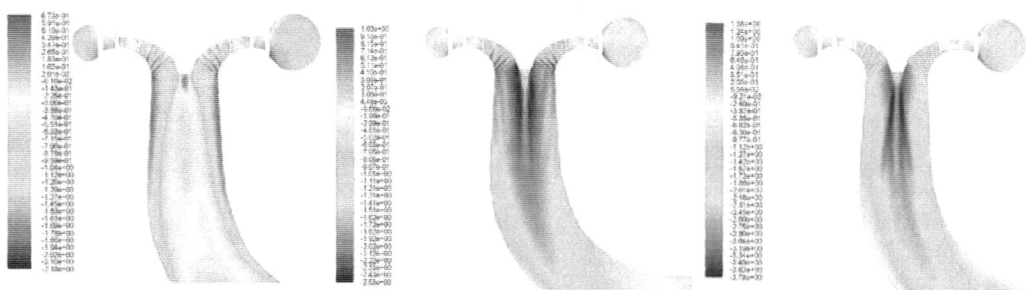

a) *b)* *c)*

Figure. 4. Isolines of the averaged axial velocity in the central section (DES model):
a) PL, b) BEP, c) HL regimes.

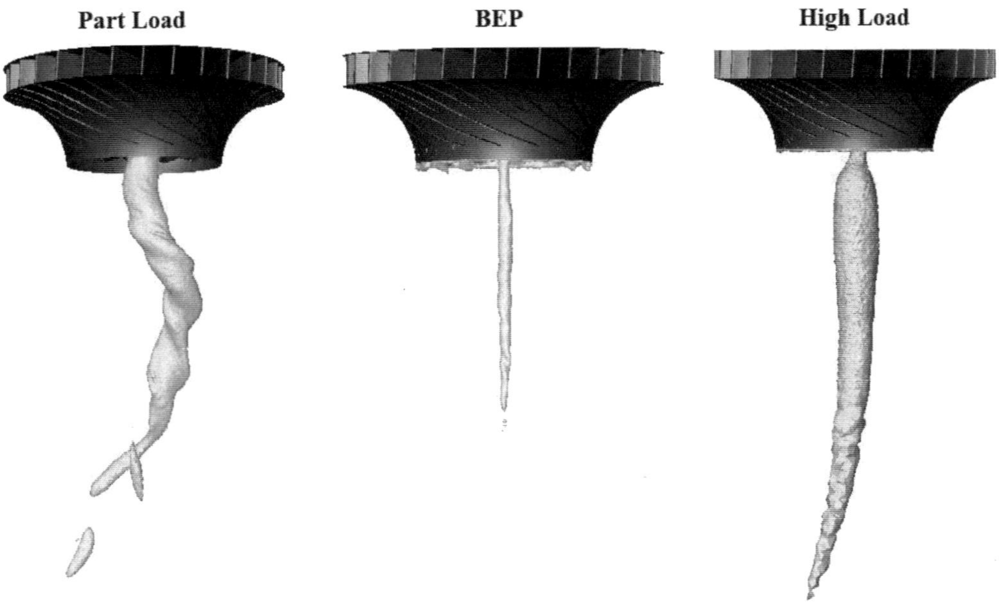

Figure. 5. Vortex structure in the draft tube for different regimes (DES model)

Comparison of the calculated and experimental radial velocity profiles are shown in Fig. 7, 9, 11. As can be seen, for all regimes the discrepancy between the calculation and the experiment is very great for all turbulence models. Perhaps, this is due to the fact, that the radial velocity has lower values compared to axial and tangential velocity and, therefore, radial velocity has high relative error of measurement.

Comparison of the calculated and experimental axial velocity profiles are shown in Fig. 6, 8, 10. As can be seen, for all regimes the calculated and experimental data agree well. The worst results are for the PL regime. For this regime all computational models give the deflection of axial velocity at the center of the draft tube. The experiment has not of this drop of axial velocity. Using of more detailed grid does not improve the result (see Fig. 6a). The results of LES calculations for the rough (5.12M) and the detailed mesh (14.28M) are similar. Also it should be noted, that the difference between the different models is not very large.

IOP Conf. Series: Journal of Physics: Conf. Series **782** (2017) 012005 doi:10.1088/1742-6596/782/1/012005

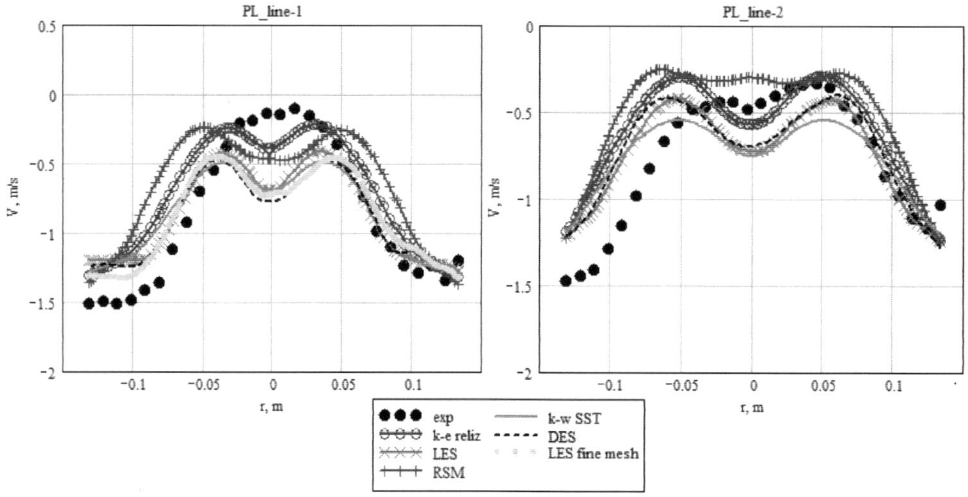

a) b)

Figure. 6. The axial component of the velocity in the line 1 (a) and line 2 (b) for PL regime.

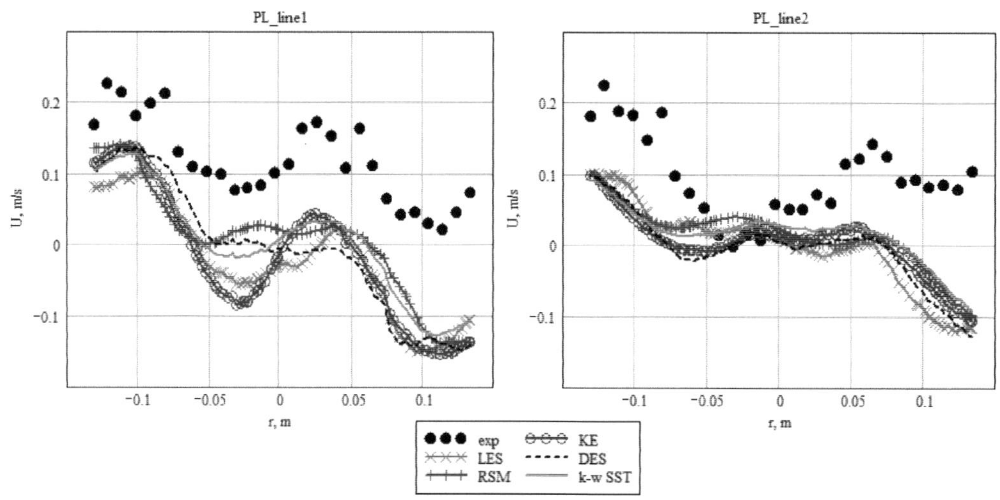

a) b)

Figure. 7. The radial component of the velocity in the line 1 (a) and line 2 (b) for PL regime.

For BEP and HL regimes the agreement between the calculation and experiment is a lot better. The best agreement between the calculation and experiment is obtained for the BEP regime. The results of the analysis of the axial velocity profiles showed that the best agreement with experiment has an LES model. The DES model also gives good results. The RANS models are close to each other. As mentioned above, the radial velocity profiles analysis does not make sense.

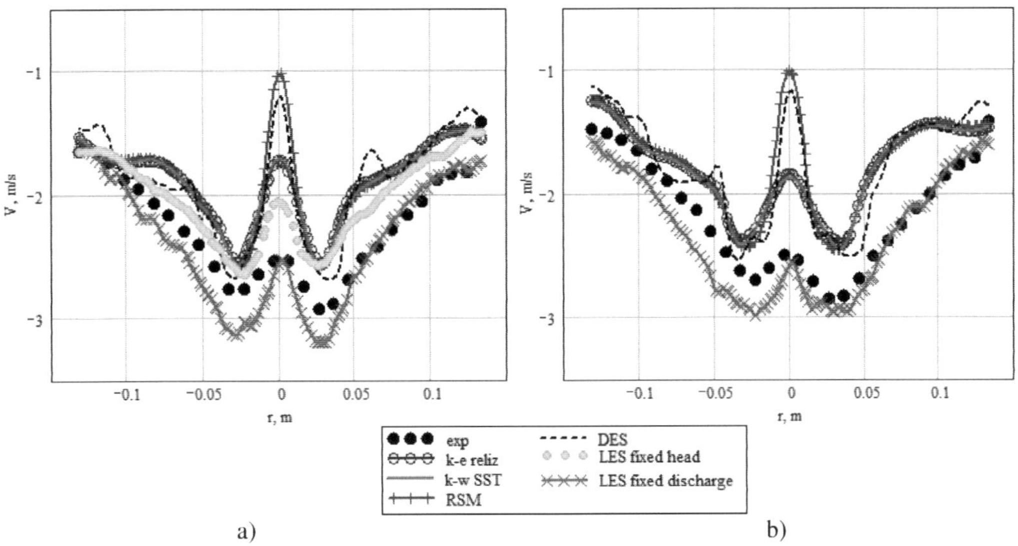

Figure. 8. The axial component of the velocity in the line 1 (a) and line 2 (b) for BEP regime.

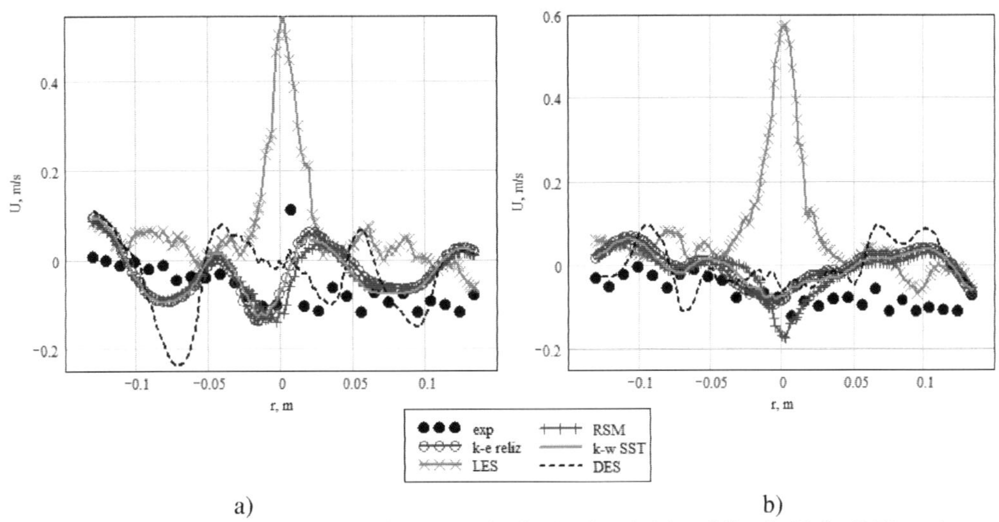

Figure. 9. The radial component of the velocity in the line 1 (a) and line 2 (b) for BEP regime.

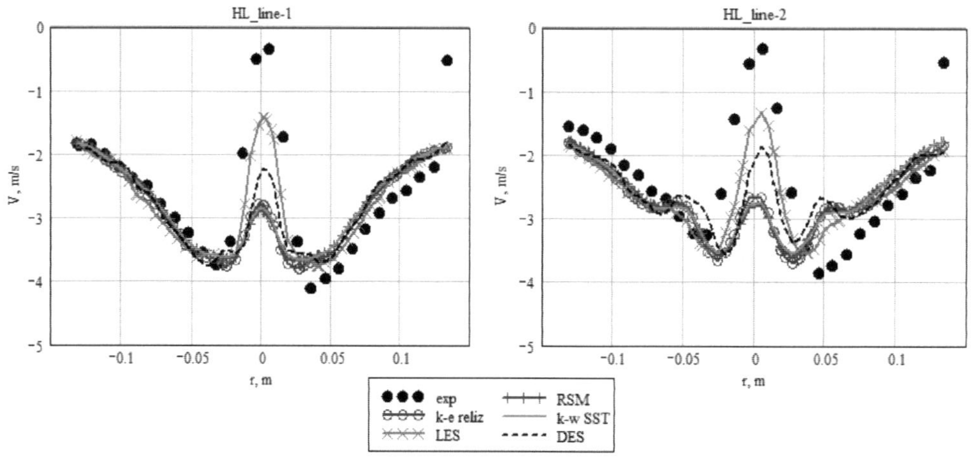

a) b)

Figure. 10. The axial component of the velocity in the line 1 (a) and line 2 (b) for HL regime.

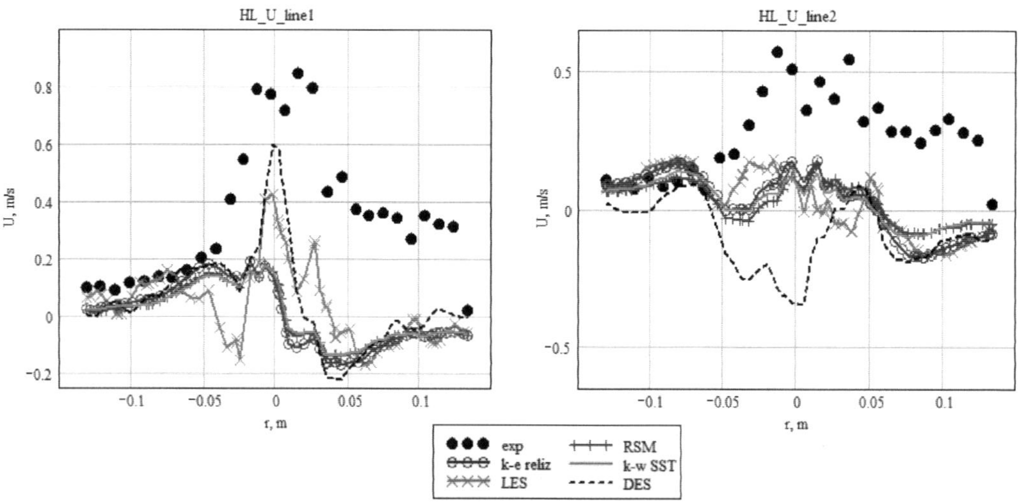

a) b)

Figure. 11. The radial component of the velocity in the line 1 (a) and line 2 (b) for HL regime.

Another reason of disagreement between calculation and experiment is different value of discharge in the calculation and experiment. At given head the calculated value of discharge is about 10% lower than the experimental value (see Table 2-4). For this reason, the calculated profiles are slightly higher than the experimental. The discrepancy between experiment and simulation is significantly reduced in case of fixing the discharge on the turbine input. An example of this is shown in the Fig.8a for LES calculation.

Comparison of the calculated and experimental integral characteristics of the turbine for all regimes are shown in the Tables 2-4. As can be seen, for all regimes calculated characteristics are 7-10% lower than the experimental value, except of the efficiency, which calculated value overestimated in comparison with the experimental one. The part load operation point shows the largest discrepancy.

The calculations by different turbulence models give similar results. It is impossible to say which model is best for finding the integral parameters.

Table 2. Part load

Parameter	k-e reliz.	SST	RSM	DES	LES	EXP.
Net head [m]	11.87	11.87	11.87	11.87	11.87	11.87
Total torque [N m]	388.8	391.5	387.5	385.5	384.8	416.39
Discharge [kg/s]	125.5	126.95	124.6	124.32	124.5	139.6
Hydraulic efficient [%]	92.9	92.5	93.3	93.03	92.7	90.13

Table 3. Best efficiency point

Parameter	k-e reliz.	SST	RSM	DES	LES	EXP.
Net head [m]	11.94	11.94	11.94	11.94	11.94	11.94
Total torque [N m]	578.2	579.01	580.11	578.7	580.5	616.16
Discharge [kg/s]	182.8	182.9	188.59	182.2	189.2	199.6
Hydraulic efficient [%]	94.2	94.9	92.3	95.3	92.1	92.39

Table 4. High efficiency point

Parameter	k-e reliz.	SST	RSM	DES	LES	EXP.
Net head [m]	11.88	11.88	11.88	11.88	11.88	11.88
Total torque [N m]	722.8	706.8	710.4	722.86	714.2	740.54
Discharge [kg/s]	229.0	226.42	226.41	228.86	231.01	242.46
Hydraulic efficient [%]	94.6	93.6	94.1	94.7	92.74	91.71

Practically, it is very important to know the level of the pressure pulsation in the turbine. Therefore, it is interesting to compare the calculated and experimental values of the pressure fluctuations. These data are shown in Fig. 12. The plot shows the static pressure on the draft tube wall. The pressure were received at the point DT5, (see workshop web site and Fig. 1).

Obtaining of the maximum level of pressure fluctuations is very important for safe operation of the turbine. Analysis of the results shows that the magnitude of the experimental pressure fluctuations is 1.68 kPa. The calculations are in good agreement with experiment (see Fig.12). Thus, the LES calculation give maximum of the pressure fluctuations 1.07 kPa, DES model give 0.99 kPa, and RSM model give 1.22 kPa. This is interesting, because it was previously thought that the RSM model should give the lower level of the pressure fluctuations. Thus it is shown that the RSM model can be used to estimate the level of the pressure pulsations in the turbines.

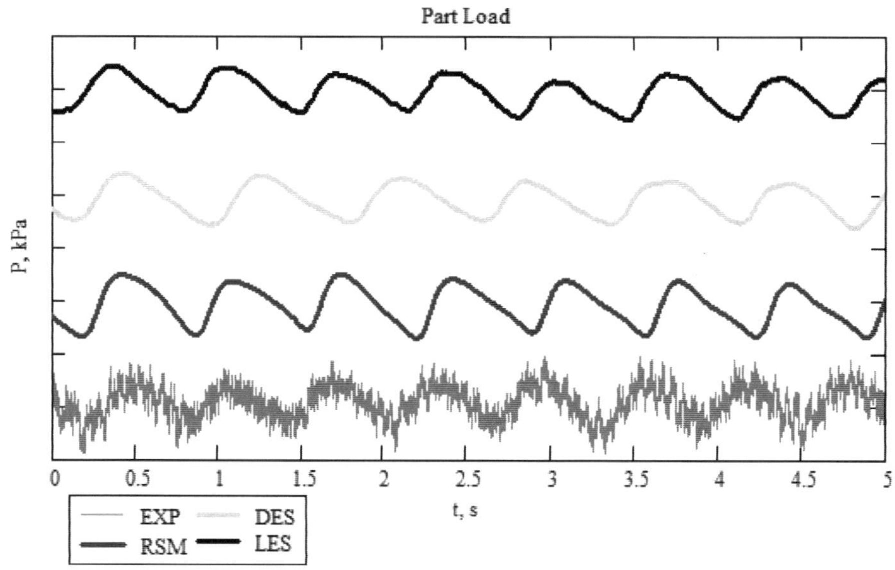

Figure. 12. The pressure pulsations at the DT5 point for Pl regime.

The behavior of the flow structure in the turbine is of great interest for analysis of the pressure pulsation of the flow. Fig. 13 shows vortex structure of the flow downstream the runner for different unsteady models at PL regime. Flow structure is shown by an instantaneous values of iso-surfaces of pressure. We see, that in this regime vortex rope structure is formed. In this case the LES and DES calculation gives two intertwined precessing vortex. Also it is interesting that, the Reynolds stress model captures well the phenomenon, although with smoothed fine-scale structures. It should be noted that the URANS model also provides some transient pressure pulsation in the draft tube.

Interesting details can be observed regarding to the pressure pulsations and axial velocity profile downstream the turbine runner. In the modes with the high pressure pulsations, the wide areas of counterflow are visible (Fig. 6). This again proves the fact that a zone of counterflow on the axial velocity profile can be an indicator of the intensity of transient phenomena in the turbine [4, 5].

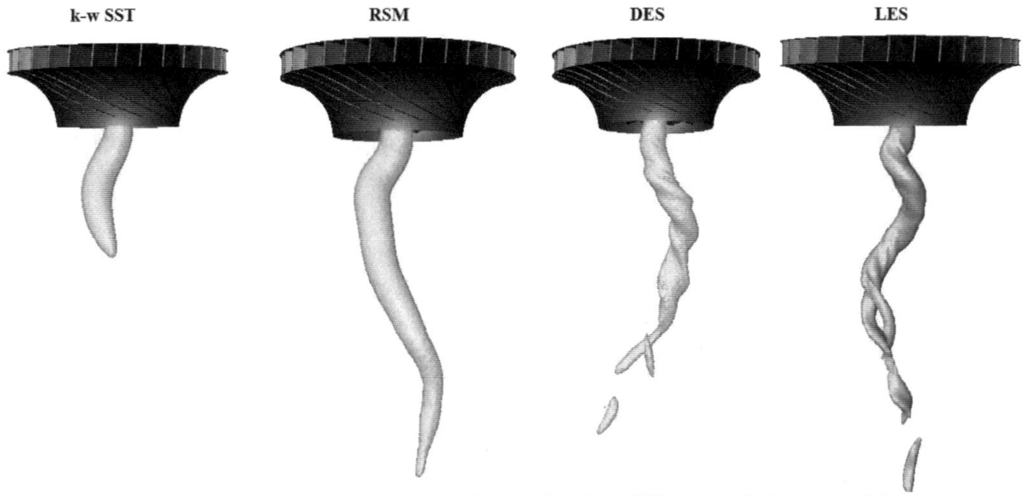

Figure. 13. Vortex structure in the PL regime for different turbulence models.

In addition to the pressure pulsation it is also very important to know the frequency of the pressure pulsations. It is important for estimating the potential resonance of the turbine design. A comparison of the experimental and calculated spectra is shown in Fig. 14. The flow after the runner has precessing vortex structures (Fig. 13). For this reason, the spectrum has a dominant frequency. In the experimental data the frequency of the pressure pulsations for this regime is 1.53Hz. This frequency is the frequency of the precession of intense vortex core.

As it is obvious from the spectrum of the pressure pulsations, the calculation results are well proved by the experiment, both for the frequency and intensity of the pulsations. The LES calculation provides frequency 1.57 Hz, DES provides frequency 1.22 Hz. And the RSM calculation provides frequency 1.47 Hz.

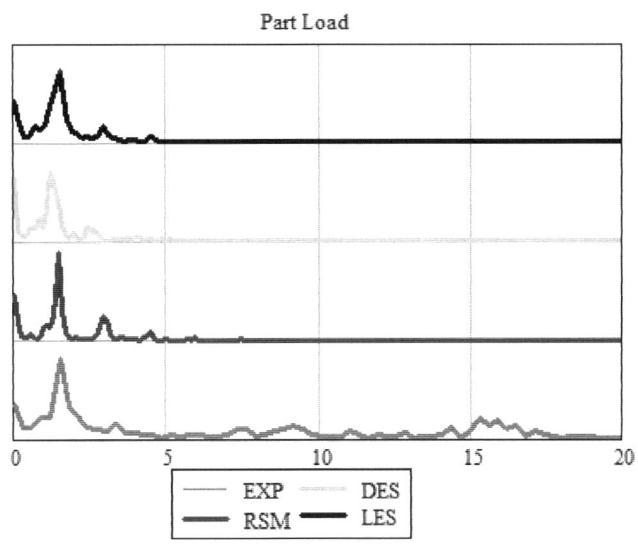

Figure. 14. The spectrum of pressure pulsations at the DT5 point for PL regime.

5. Conclusions

The paper presents the results of numerical simulation of a flow in a laboratory model of a Francis hydroturbine at three regimes, using two eddy-viscosity- (EVM) and a Reynolds stress (RSM) RANS models (realizable k-ε, k-ω SST, LRR) and detached-eddy-simulation (DES), as well as large-eddy simulation (LES). Three operation point proposed for the Francis-99 workshop were calculated. Comparison of calculation results with the experimental data was carried out.

The results of the analysis of the axial velocity profiles showed that the best agreement with experiment has an LES model. The DES model also gives good results. The RANS models are close to each other. The worst results are for the PL regime. For BEP and HL regimes the agreement between the calculation and experiment is a lot better. The best agreement between the calculation and experiment is obtained for the BEP regime.

The reason of disagreement between calculation and experiment may be different value of discharge in the calculation and experiment. At given head the calculated value of discharge is about 10% lower than the experimental value (see Table 2-4). For this reason, the calculated profiles are slightly higher than the experimental. The discrepancy between experiment and simulation is significantly reduced in case of fixing the discharge on the turbine input.

The structure of the flow behind the runner was analyzed by numerical simulation, which revealed the effect of the flow structures on the frequency and intensity of the pulsations. The calculations confirmed that a vortex rope behind the runner is the main cause of low-frequency pressure pulsations in hydraulic turbines. Its dynamics determines predominantly the behavior of turbine pulsations.

Acknowledgement
This work was supported in part by the Russian Foundation for Basic Research.

References

[1] Nicolet C 2007 Hydroacoustic modelling and numerical simulation of unsteady operation of hydroelectric systems *PhD Thesis* (EPFL № 3751, http://library.epfl.ch/theses/?nr=3751).

[2] Doerfler P 1982 System dynamics of the Francis turbine half load surge. *IAHR Symp. on Hydraulic Machinery and Systems, Amsterdam, Netherlands.*

[3] Ruprecht A 2006 Numerical prediction of vortex instabilities in turbomachinery *Notes on Numerical Fluid Mechanics and Multidisciplinary Design* **93** p. 211-224.

[4] Minakov A V, Platonov D V, Dekterev A A, Sentyabov A V and Zakharov A V 2015 The analysis of unsteady flow structure and low frequency pressure pulsations in the high-head Francis turbines *International Journal of Heat and Fluid Flow* **53** pp. 183 – 194

[5] Minakov A V, Platonov D V, Dekterev A A, Sentyabov A V and Zakharov A V 2015 The numerical simulation of low frequency pressure pulsations in the high-head Francis turbine *Computers and Fluids* **111, April** pp. 197 – 205. doi:10.1016/j.compfluid.2015.01.007

[6] Alligne S, Maruzewski P, Dinh T, Wang B, Fedorov A, Iosfin J and Avellan F 2010 Prediction of a Francis turbine prototype full load instability from investigations on the reduced scale model. *Proc. IAHR Symp. on Hydraulic Machinery and Systems, Timisoara, Romania.*

[7] Cervantes M J, Engstrom T F, Gustavsson L H 2005 *Proc. The third IAHR/ERCOFTAC workshop on draft tube flows Turbine 99. Sweden, Porjus.*

[8] Minakov A V, Sentyabov A V, Platonov D V, Dekterev A A, Gavrilov A A 2014 Numerical modeling of flow in the Francis-99 turbine with Reynolds stress model and detached eddy simulation method *Journal of Physics: Conference Series* **579**, doi:10.1088/1742-6596/579/1/012004.

[9] Gavrilov A, Dekterev A, Sentyabov A, Minakov A, Platonov D 2012 Application of hybrid methods to calculations of vortex precession in swirling flows *Notes on Numerical Fluid Mechanics* **117** pp. 449-459.

[10] Kuznetsov I, Zakharov A, Orekhov G, Minakov A, Dekterev A, Platonov D 2012 Investigation of free discharge through the hydro units of high head Francis turbine. *IOP Conference Series: Earth and Environmental Science* **15 (PART 5)**. doi: 10.1088/1755-1315/15/5/052002

[11] Smirnov R, Shi S, Celik I 2001 Random flow generation technique for large-eddy simulations and particle-dynamics modeling *J. Fluids Engineering*. **123** pp. 359–371.
[12] Picano S and Hanjalić K 2012 Leray-α regularization of the Smagorinsky-closed filtered equations for turbulent jets at high Reynolds numbers *Flow, Turbulence and Combustion*, **89**: 627–650.

IOP Conf. Series: Journal of Physics: Conf. Series **782** (2017) 012006 doi:10.1088/1742-6596/782/1/012006

Steady state operation simulation of the Francis-99 turbine by means of advanced turbulence models

A Gavrilov[1,2,3], A Dekterev[1,2,3] , A Minakov[1,2,3], D Platonov[1,2,3] and A Sentyabov[1,2,3]

[1] Siberian Federal University, Krasnoyarsk, Svobodniy 79, 660079 Russia
[2] Institute of Thermophysics SB RAS, Novosibirsk, Lavrentyeva 1, 630090 Russia
[3] Novosibirsk State University, Novosibirsk, Pirogova 2, 630090 Russia

E-mail: gavand@yandex.ru, dekterev@mail.ru, tov-andrey@yandex.ru, platonov-08@yandex.ru, sentyabov_a_v@mail.ru

Abstract. The paper presents numerical simulation of the flow in hydraulic turbine based on the experimental data of the II Francis-99 workshop. The calculation domain includes the wicket gate, runner and draft tube with rotating reference frame for the runner zone. Different turbulence models such as k-ω SST, ζ-f and RSM were considered. The calculations were performed by means of in-house CFD code SigmaFlow. The numerical simulation for part load, high load and best efficiency operation points were performed.

1. Introduction

Hydropower resources are the major renewable source of the electric power. At the present time, hydraulic power plants are being built and modernized. Accordingly, many scientific and engineering problems arise in the field of flow dynamics, such as finding of the power and efficiency of the turbine, calculation of the cavitation, pressure pulsations and forces on the turbine equipment.

Numerical simulation of flow in turbines is a useful tool for design of hydraulic turbines. At the same time, this is a very complicated problem due to complex shape of the flow path, turbulence, rotated runner, swirled flow under the runner and large-scale unsteady coherent structures. These features impose strict requirements on the turbulence models. Two-equation turbulence models (k-ε and k-ω) are most widely used in industrial applications. Large eddy simulation (LES) and hybrid RANS/LES methods are increasingly used for unsteady large-scale phenomena in complicated flows. At last, differential Reynolds stress model can accurately reproduce phenomena related to anisotropy of the turbulent fluctuations.

Several international projects were conducted for investigating flow in the hydraulic turbines with numerical simulation. The flow in the Kaplan turbine draft tube was considered in three Turbine-99 workshops. The Turbine-99 workshop organizing committee provide geometry of the draft tube, the experimental inlet velocity profiles and the other experimental data. Scientific groups performed a lot of steady and unsteady numerical simulations by means of different turbulence models: LES [1], DES [2], unsteady Spalart-Allmaras model [3] etc. However, the calculations strongly depends on the inlet conditions [4]. Another project of the turbine simulation was FLINDT. For example, in [5] numerical simulations of the rotating vortex were performed with the k-ε turbulence model in the compromise computational domain that includes runner and draft tube. A number of unsteady approaches (URANS, DES, DDES and so on) were considered for numerical simulation of the flow in a test rig with stationary vertical guide vanes and a free runner [6].

Francis-99 is a project in the same spirit as the Turbine-99. It is a series of three workshops, which provides an open access of the complete design and data of a model Francis turbine [7]. Many scientist groups can use these data for validation of the different CFD codes and numerical approach. First workshop was in 2014 and 14 papers were presented. Those simulations were performed with *k*-ε, *k*-ω SST, SAS, RSM, zonal LES turbulence model and *k*-ω SST was the most widely used.

Previously, flow in the Francis-99 draft tube were simulated by means of the in-house CFD code SigmaFlow for the experimental data of the first workshop. Now, this paper presents an attempt of numerical simulation of the flow in the full turbine, including draft tube, runner and guide vanes, by means of the code SigmaFlow. The rotated reference frame approach validated for the code. Different turbulence models were considered for three operating points as well.

2. Mathematical model and numerical method

The simulations were based on the geometry and parameters of the operating points provided by organizers of the Francis-99 workshop II [7]. Computational domain consists of wicket gate, runner and draft tube (Fig. 1). Rotated reference frame in absolute velocity formulation was used for modeling of the runner rotation.

Below, the basic equations of the mathematical models expressing the conservation laws in the rotating reference frame are presented. Einstein summation convention is used.

The continuity equation (conservation of mass):

$$\frac{\partial u_i}{\partial x_i} = 0 \qquad (1)$$

Momentum equations (conservation of momentum) in a rotating reference frame for absolute velocities:

$$\frac{\partial \rho u_i}{\partial t} + \frac{\partial}{\partial x_j}\left(\rho u_j^r u_i\right) = -\frac{\partial p}{\partial x_i} + \frac{\partial}{\partial x_j}(\tau_{ij}^{\mathrm{m}} + \tau_{ij}^{\mathrm{t}}) - \rho \varepsilon_{ijk} \Omega_j u_k \qquad (2)$$

Where: u_i – absolute velocity components, u_j^r – relative velocity components, τ_{ij}^{m} – viscous stress tensor, τ_{ij}^{t} – turbulent stress tensor, Ω – angular velocity of runner rotation, p – static pressure, ρ – density, ε_{ijk} – Levi-Civita symbol. Several turbulence modeling approach were used. There were eddy viscosity models (*k*-ω SST, [8, 9], *ζ-f* [10]) and differential Reynolds stress model [11]. Pressure strain term is modeled by SSG strain model [11].

Due to the poor mesh, the computational domain did not include spiral case, therefore the head can not be used as a boundary condition. The discharge, obtained from the experimental data, was fixed at the inlet and outlet boundaries. The inlet velocity was directed normal to the wicket gate inlet. Speed of the runner rotation was -34.83 rad/s according to experimental parameters and the runner wall rotated with corresponding velocity. The density of the water was $\rho = 999.8$ kg/m^3 and the molecular viscosity was $\mu = 9.568 \cdot 10^{-4}$ Pa·s.

Computational meshes includes about 4.3 mln. polyhedral cells (see Table 1, Fig. 1, b). Dimensionless wall distance varied from 80 in the runner and draft tube cone up to 10^3 in the draft tube tail. Standard wall function was used in case of eddy viscosity model and a blended technique [13] was used in case of RSM model.

Table 1. Mesh resolution

	Cells, mln. cells
Wicket gate	0.57
Runner	1.76
Draft tube	1.94

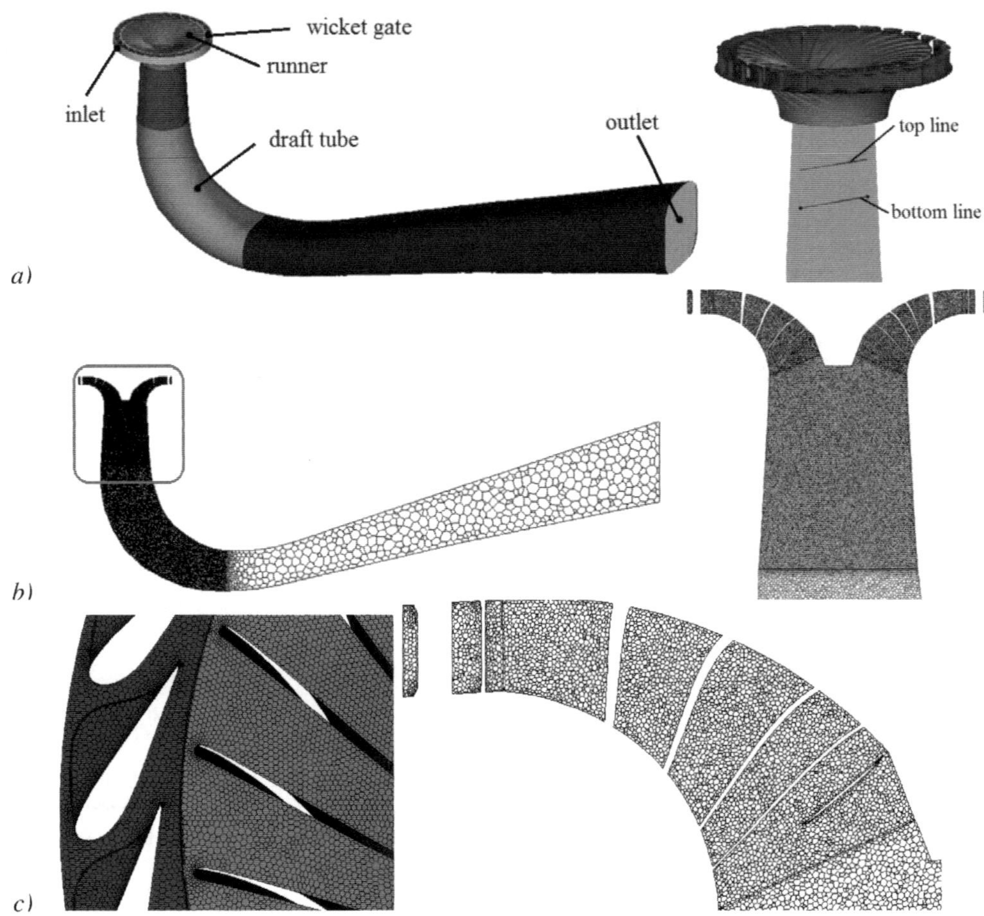

Figure 1. Francis-99 test case; *a)* computational domain; *b)* mesh in central cross-section, c) detailed views of the mesh on the guide vanes and the runner

The simulations were performed by means of in-house CFD code SigmaFlow [12, 13] which is developed in the Institute of thermophysics, Novosibirsk. The code was designed for solving of hydrodynamics and thermophysics problems in industrial applications and scientific investigations in the field of numerical simulation of turbulence, cavitation, non-Newtonian flow and heat transfer. Discretization of transport equations was carried out by the control volume method on unstructured grids which can includes both hexahedral and polyhedral cells. Program package includes mesher and can use imported meshes. Analysis of the solution is performed by means of the built-in 3D visualizer. Parallel calculations are performed by means of domain decomposition and MPI.

Coupling of the velocity and pressure fields for incompressible flow was realized using the multistep SIMPLE-like procedure. Schemes of different order are available for the approximation of the convective terms in the equations for momentum components but in these calculation only up-wind scheme was used due to poor robustness. For the approximation of the convective terms in the equations for turbulent characteristics up-wind scheme was used as well. Unsteady calculation were performed to obtain steady or averaged solution. Time step was 0.01 s in case of k-ω SST and ζ-f turbulence models and 10^{-4} s in case of Reynolds stress model since RSM had worse convergence.

Earlier some calculations have shown that this technique can reliably consider averaged velocity and large-scale turbulent fluctuations in water turbines [8–9].

3. Results of Francis-99 simulations

In case of fixed discharge, the calculated torque on the runner is overestimated noticeably in the considered operating points (Table 2). Models k-ω SST and ζ-f calculates the same value of the torque, but RSM model obtains the lesser one. The discrepancies of the torque vary from 9 to 16%. Since the calculated torque is larger than experimental one the discrepancies is not seem a results of the mesh resolution. Fig. 2 shows that there is small difference of pressure distribution on the runner blades between k-ω SST and ζ-f results. Reynolds stress models demonstrates the same results near the leading edges of the blades but the flow around trailing edges is different (see Fig. 2, 3). In case of RSM, the region of the low pressure is localized on a small area near the leading edge (Fig. 3).

Table 2. Torque on the runner, N·m (discrepancies, %)

	PL	BEP	HL
k-ω SST	487 (16%)	703 (13%)	835 (12%)
ζ-f	486 (15%)	702 (13%)	833 (12%)
RSM	465 (10%)	680 (9%)	810 (9%)
experiment	421	621	744

At the best efficiency operating point, the flow under the runner is steady and weakly swirled (Fig 4*b*, 5*b*). There is small recirculation region under the runner hub. This zone generates long wake along the turbine axis. There is a straight weak vortex at the turbine axis. The axial velocity profiles have the maximum around the wake (Fig 4*b*, 6*b*). The calculated axial velocity profiles closely agree with the experimental profiles. There is no graphical differences between results of the eddy viscosity models (k-ω SST and ζ-f).

At the part load operating point, large-scale unsteady vortex rope forms under the runner (Fig 4*a*, 5*a*). There are many small vortices under the runner blades near the draft tube wall (Fig 5*a*). Calculated axial velocity profiles show a recirculation zone near the axis, but experimental profiles do not show it (Fig. 6*a*). Therefore, the calculated vertical velocity is significantly above the experimental one at the axis in case of all the models.

At the high load operating point, there is a weak vortex under the runner hub (Fig 4*c*, 5*c*). The vortex undergoes instability and has a spiral form at a distant of the runner. Calculated axial velocity profiles agree with experimental data but they show significant numerical viscosity for all the models (Fig. 6*c*).

IOP Conf. Series: Journal of Physics: Conf. Series **782** (2017) 012006 doi:10.1088/1742-6596/782/1/012006

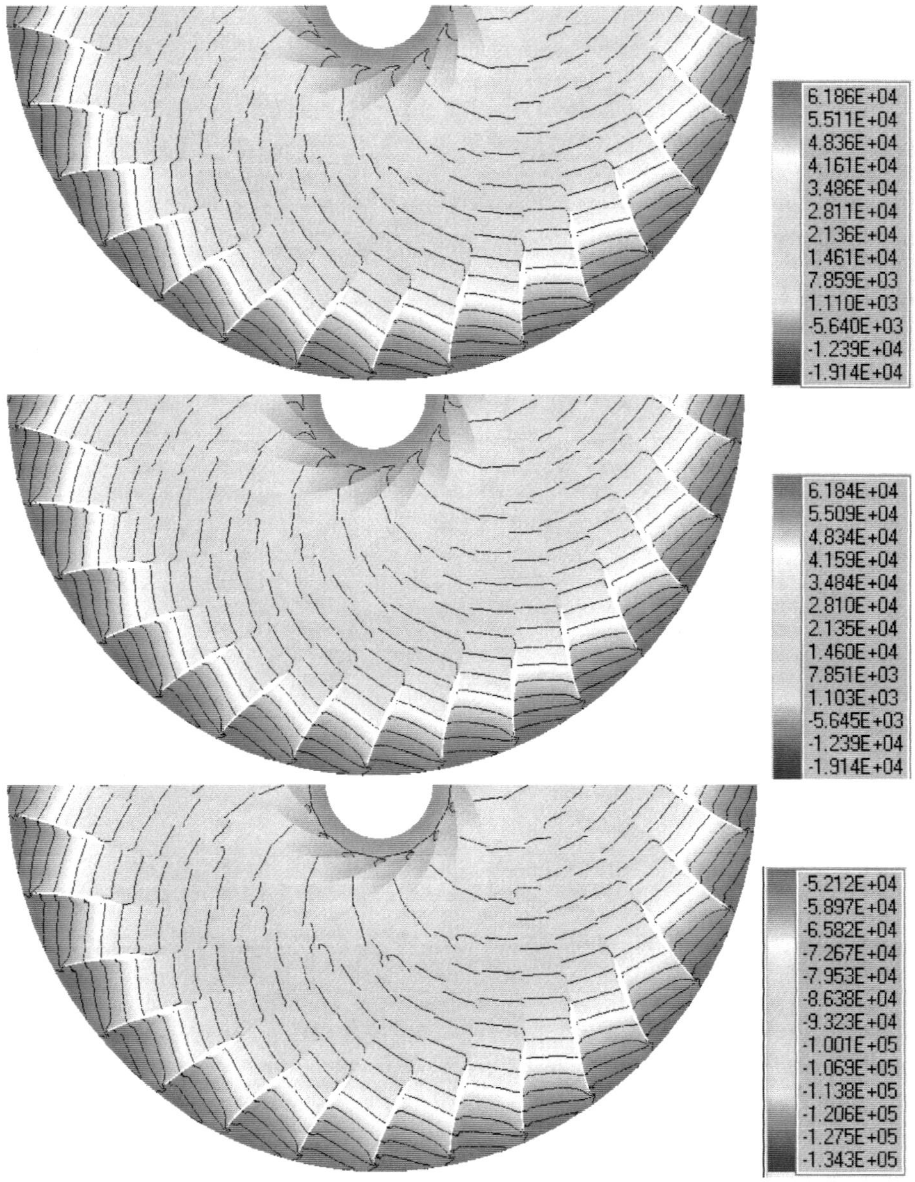

Figure 2. Pressure on the runner at the best efficiency point: *a*) *k*-ω SST; *b*) ζ-*f* , *c*) RSM

IOP Conf. Series: Journal of Physics: Conf. Series **782** (2017) 012006 doi:10.1088/1742-6596/782/1/012006

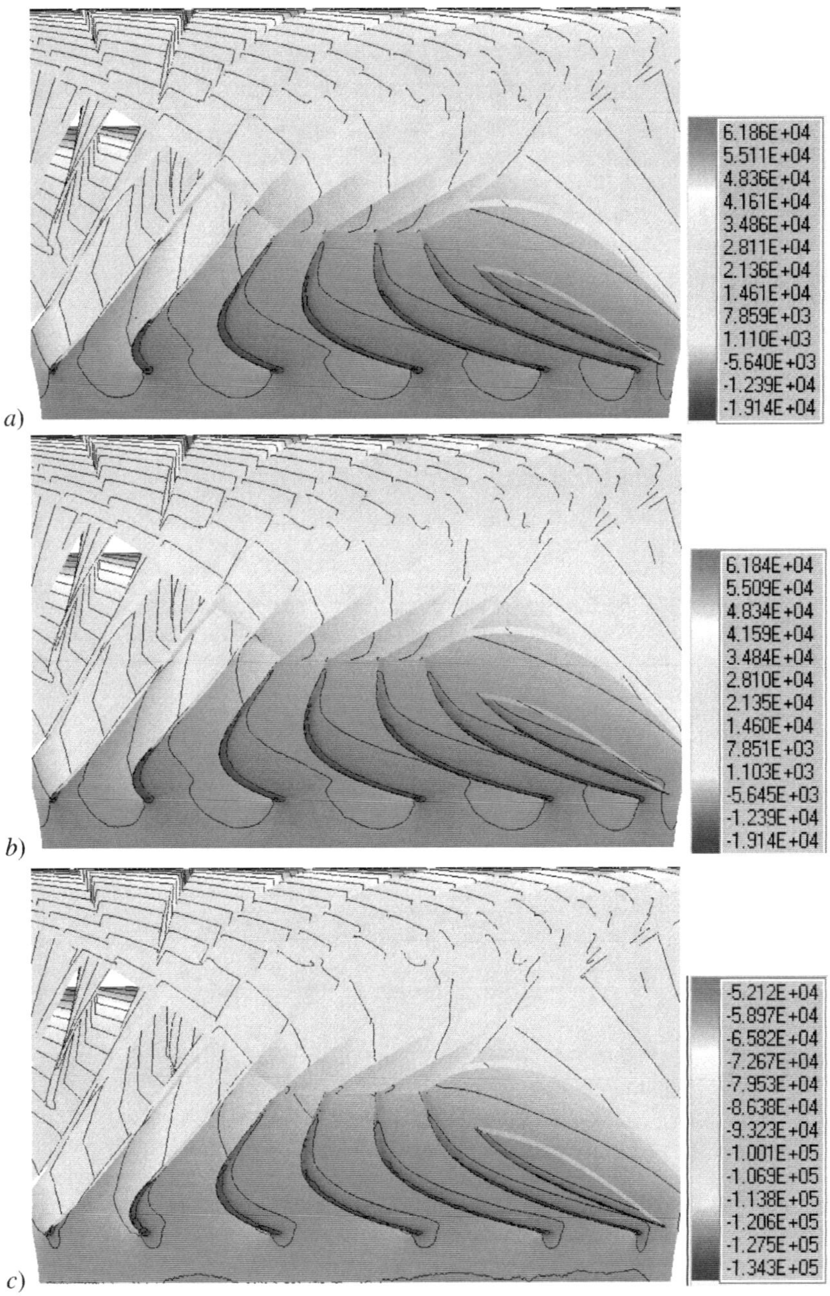

Figure 3. Pressure on the trailing edge of the runner blades at the best efficiency point: *a*) *k*-ω SST; *b*) ζ-*f*, *c*) RSM

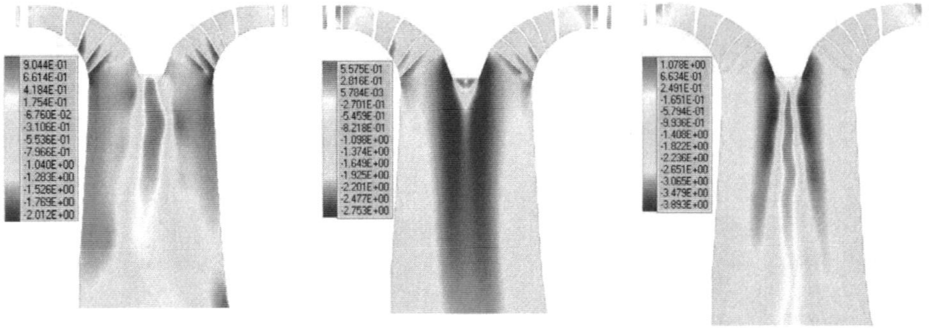

a *b* *c*

Figure 4. Instant vertical velocity in the central plane (ζ-*f* model): *a*) part load, *b*) best efficiency point, *c*) high load.

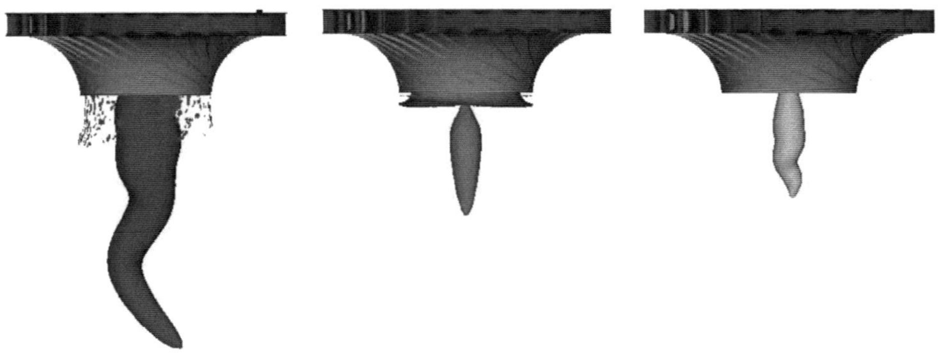

a *b* *c*

Figure 5. Vortices in the draft tube, visualized by iso-pressure surface (ζ-*f* model): *a*) part load, *b*) best efficiency point, *c*) high load.

IOP Conf. Series: Journal of Physics: Conf. Series **782** (2017) 012006 doi:10.1088/1742-6596/782/1/012006

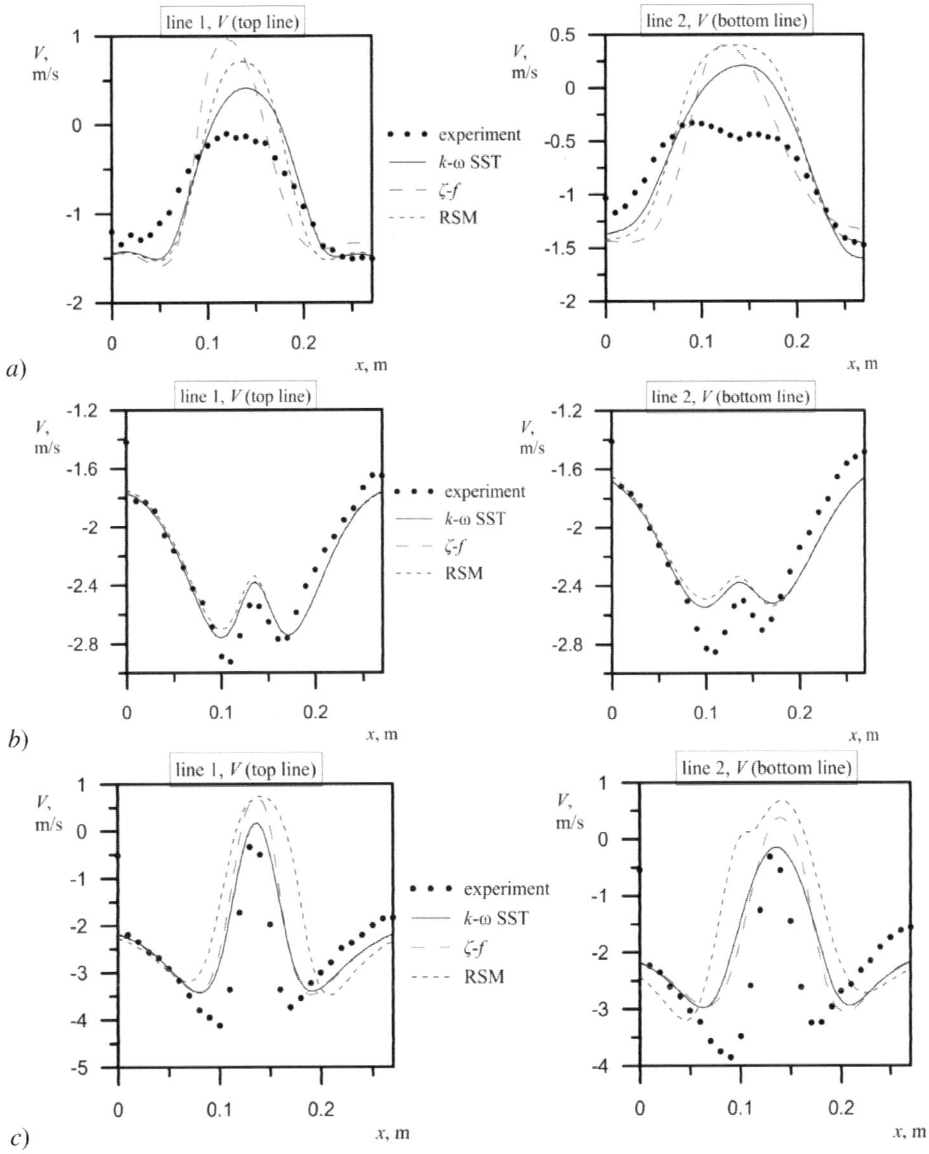

Figure 6. Averaged axial velocity component in the draft tube: *a*) part load; *b*) best efficiency point, *c*) high load.

4. Conclusions

Thus, this paper presents the first example of turbine simulation by means of the code SigmaFow in such formulation (wicket gate, rotor and draft tube with rotated reference frame for the runner zone). This kind of calculation is a challenge for the computational resources. For example, eddy viscosity simulations takes several days computation by means of 24 CPU cores.

The calculations show flow patterns depending on the operating point of the turbine. At the best efficiency point the flow in the draft tube is steady and there is a weak straight vortex along the turbine axis under the runner. At part load operating point there is a wide strong vortex rope that rotates near the draft tube wall. At high load operating point there is a weak vortex under the runner hub which form becomes spiral at a distant of the runner.

Three modern turbulence models were considered. k-ω SST model is one of the most widely used in industrial applications and robust model. ζ-f model is an advanced four-equation model that can accurately describe near wall turbulence. Differential Reynolds stress model is a promising tool in case of complex, especially swirled, turbulence flow. There are still many unresolved problems off Reynolds stress model related to wall function, robustness, computational efficiency and so on.

The calculated data agree with experimental results well for best efficiency point. For high load operating points there is only qualitative agreement between calculation and experimental results. For part load operating points there is significant discrepancy between calculated and experimental results. For all the regimes there is no noticeable difference between the results of the eddy viscosity models k-ω SST and ζ-f.

Acknowledgements

This work was supported by Russian Science Foundation (project No. 16-19-00138).

References

[1] Kurosava S and Nakamura K 2005 Unsteady turbulent flow simulation in T99 draft tube *Proceedings of the third IAHR/ERCOFTAC workshop on draft tube flows Turbine 99* pp. 73–82.

[2] Marjavaara D.B., Kamakoti R., Lundstrom T.S.,Thakur S., Wright J. and Shyy W. 2005 Steady and unsteady CFD simulation of the Turbine-99 draft tube using CFX-5 and Stream *Proceedings of the third IAHR/ERCOFTAC workshop on draft tube flows Turbine 99* pp. 83–99.

[3] Page M., Giroux A.M. and Nicolle J. 2005 Steady and unsteady computation of Turbine-99 draft tube *Proceedings of the third IAHR/ERCOFTAC workshop on draft tube flows Turbine 99* pp. 109–124.

[4] Nilsson H. 2006 Evaluation of OpenFOAM for CFD of turbulent flow in water turbines *Proceedings of 23rd IAHR Symposium*

[5] Ciocan G.D. Iliescu M.S., Vu T.C., Nennemann B. and Avellan F. 2007 Experimental Study and Numerical Simulation of the FLINDT Draft Tube Rotating *Transactions of the ASME* **129** pp. 146 – 158

[6] Javadi A and Nilsson H 2015 Time-accurate numerical simulations of swirling flow with rotor-stator interaction *Flow, Turbulence and Combustion* **95(4)** pp 755 – 754 DOI: 10.1007/s10494-015-9632-2

[7] https://www.ntnu.edu/nvks/test-case.

[8] Menter F R 1994 Two equation eddy viscosity turbulence models for engineering applications *AIAA J.* **32(8)** pp 1598-1605

[9] Menter F R, Kuntz M and Langtry R 2003 Ten years of experience with the SST turbulence model *Heat and Mass Transfer* **4** pp 625–632

[10] Hanjalić K, Popovac M and Hadžiabdić M 2004 A obust near-wall elliptic relaxation eddy viscosity turbulence model for CFD *Int. J. Heat Fluid Flow* **25(6)** pp 1047-1051.

[11] Speziale C G, Sarkar S and Gatski TB 1991 Modeling the pressure–strain correlation of turbulence: an invariant dynamical system approach *J. Fluid Mech.* **227** pp 245-272.

[12] Gavrilov A A, Dekterev A A and Sentyabov A V 2015 Modeling of swirling flows with coherent structures using the unsteady reynolds stress transport model *Fluid Dynamics* Vol. **50(4)** pp 471 – 482.

[13] Gavrilov A A, Sentyabov A V, Dekterev A A and Hanjalic K 2016 Vortical structures and pressure pulsations in draft tube of a Francis-99 turbine at part load: RANS and hybrid

RANS/LES analysis *International Journal of Heat and Fluid Flow* **June** (online available), DOI: 10.1016/j.ijheatfluidflow.2016.05.007

[14] Minakov A V, Platonov D V, Dekterev A A, Sentyabov A V and Zakharov A V 2015 The numerical simulation of low frequency pressure pulsations in the high-head Francis turbine *Computers and Fluids* **111** pp 197-205, DOI:10.1016/j.compfluid.2015.01.007
Minakov A V, Platonov D V, Dekterev A A, Sentyabov A V and Zakharov A V 2015 The analysis of unsteady flow structure and low frequency pressure pulsations in the high-head Francis turbines *International Journal of Heat and Fluid Flow* **53** pp 183-194, DOI: 10.1016/j.ijheatfluidflow.2015.04.001

IOP Conf. Series: Journal of Physics: Conf. Series **782** (2017) 012007 doi:10.1088/1742-6596/782/1/012007

Performance prediction of the high head Francis-99 turbine for steady operation points

E. Casartelli, L. Mangani, O. Ryan, A. Del Rio

Lucerne University of Applied Science & Arts, Department of Mechanical Engineering, Horw, Switzerland

E-mail: ernesto.casartelli@hslu.ch

Abstract. Steady-state numerical investigations are still the reference computational method for the prediction of the global machine performance during the design phase. Accordingly, steady state CFD simulations of the complete high head Francis-99 turbine, from spiral casing to draft tube have been performed at three operating conditions, namely at part load (PL), best efficiency point (BEP), and high load (HL). In addition, simulations with a moving runner for the three operating points are conducted and compared to the steady state results. The prediction accuracy of the numerical results is assessed comparing global and local data to the available experimental results. A full 360°-model is applied for the unsteady simulations and for the steady state simulations a reduced domain was used for the periodic components, with respectively only one guide vane and one runner passage. The steady state rotor-stator interactions were modeled with a mixing-plane. All CFD simulations were performed at model scale with an in-house 3D, unstructured, object-oriented finite volume code designed to solve incompressible RANS-Equations. Steady and unsteady solver simulations are both able to predict similar values for torque and head in design and off-design. Flow features in off-design operation such as a vortex rope in PL operation can be predicted by both simulation types, though all simulations tend to overestimate head and torque. Differences among steady and unsteady simulations can mainly be attributed to the averaging process used in the mixing plane interface in steady state simulations. Measured efficiency agrees best with the unsteady simulations for BEP and PL operation, though the steady state simulations also provide a cost-effective alternative with comparable accuracy.

1. Introduction

Accurate and trustworthy numerical predictions of turbine performance in hydropower generation are more important than ever in the current competitive energy production environment. New developments in CFD codes provide quick, accurate and improved flow field predictions for hydropower applications. The Francis 99 workshop presents an opportunity for academic and industrial researchers to engage in research of the Tokke high head turbine model. After the first successful workshop in 2014, this second workshop allows to undertake more in-depth research and with an extended range of tasks for the participants of this event [10].

In this study, steady state and unsteady simulations are performed on the turbine in PL and high-load operating points as well as at design conditions. The computational domain reaches from the spiral casing inlet to the draft tube outlet, where a full 360°-model is used for the unsteady simulations and a passage model for the steady state computations. An in-house developed code is used here which is a block coupled incompressible solver embedded in the OpenFOAM framework [7]. Simulation results are compared to measurements which are provided, together with the computational grids. For the steady state simulations using the passage model, a fully implicit mixing plane is applied for the rotor-stator interfaces [4]. The main focus of this study is to evaluate the performance of the steady state coupled solver used here in combination with the mixing plane, which promises to be an efficient method for hydro turbine investigations and can achieve considerable speed-up compared to conventional segregated solvers [1]. The unsteady simulations serve as numerical reference cases which require considerably more computational effort.

2. CFD-solver technology

2.1. Governing equation

The simulations are based on the Unsteady Reynolds Averaged Navier-Stokes (URANS) equations, which govern the dynamics of incompressible, viscous flows. The differential form can be written in an Arbitrary Lagrangian Eulerian (ALE) framework as follows:

$$\nabla \cdot \mathbf{u} = 0 \tag{1}$$

$$\frac{\partial \mathbf{u}}{\partial t} + \nabla \cdot (\mathbf{u}_{\mathrm{r}} \mathbf{u}) = -\frac{1}{\rho} \nabla p + \nabla \cdot (\nu_{\mathrm{eff}} \nabla \mathbf{u}) \tag{2}$$

where $\mathbf{u}$ is the absolute velocity, p the pressure and ρ the density of the fluid. The convecting flux is written in terms of the relative velocity $\mathbf{u}_{\mathrm{r}} = \mathbf{u} - \mathbf{u}_{\mathrm{ALE}}$, where $\mathbf{u}_{\mathrm{ALE}}$ is the velocity associated to the dynamic motion (either rigid and/or deformable) of the physical domain $V = V(t)$ and it is evaluated enforcing the so-called Geometric Conservation Law (GCL). For instance, for a purely rigid rotation with constant angular velocity $\mathbf{\Omega}$, it corresponds to $\mathbf{u}_{\mathrm{ALE}} = \mathbf{\Omega} \times \mathbf{r}$. Finally the effective kinematic viscosity is the sum of the laminar and the turbulent contributions ($\mu_{\mathrm{eff}} = \mu + \mu_{\mathrm{t}}$).

2.2. Numerical discretization

All simulations were performed using a so called coupled solver. The difference to the conventional segregated approach lies in the way the governing equations are solved. Speed and robustness are among the most important requirements for any software that has to be used for the time-accurate numerical analysis of complex highly unsteady phenomena. In the coupled solver the governing equations are coupled implicitly, since resolving efficiently the pressure-velocity coupling is essential for the performance of any CFD code. In comparison to the segregated solver of the SIMPLE family of algorithms, which sequentially solve each governing equation, the coupled approach benefits of higher robustness and speed. The coupled solver used for the simulations bases on the open-source CFD library OpenFOAM and was developed by Mangani et al. [7].

2.3. Turbulence modeling

All the simulations were performed using a SST k-ω turbulence model with automatic wall treatment. The automatic wall treatment decides whether the boundary layer is being resolved or a wall-function is used. This type of turbulence model requires two auxiliary scalar equations (not coupled) in addition to the momentum and continuity equation, to compute the conservation variables of turbulent kinetic energy k and specific dissipation rate ω.

Although two equation turbulence models are known to have shortcomings at off-design conditions, it has been shown in the first workshop of Francis 99 [9], that such a model is capable to produce suitable results compared with the measurements, even for operation apart from the best efficiency point [1]. The choice of the turbulence model is based also on the fact that it is widely accepted in industrial applications [12] and corresponds to the philosophy of applied science.

IOP Conf. Series: Journal of Physics: Conf. Series **782** (2017) 012007 doi:10.1088/1742-6596/782/1/012007

2.4. Domain discretization

For the steady state and the unsteady simulation of static operating points, two different meshes were applied. For the unsteady simulations a complete 360°-model is necessary to capture the unsteady flow-behavior. The simulation domain ranges from the spiral case inlet, the stay vanes, the guide vanes and runner to the draft tube as shown in fig.1 (a). Rotating and stationary domains are matched using a rotor stator arbitrary mesh interface (AMI).

For the steady state simulations, a single passage for the guide vane and the runner is applied, see fig.1 (b). In this case the flow-parameters are interpolated between the stay vanes and the guide vane passage, between the guide vane passage and the runner passage and also between the runner passage and the draft tube. The interpolation is performed using the mixing plane method. For the simulation a fully implicit mixing plane was used [4]. Therefore, flow-parameters are averaged in the circumferential direction. This method bases on the observation, that flow information in circumferential direction mixes out faster than in Z- direction (for the interfaces between stay vanes and runner) or r-direction (for the interface between runner and draft tube).

The computational grids for both steady state and unsteady models are provided by the workshop committee. A mesh study is not performed here. Three operating points at different

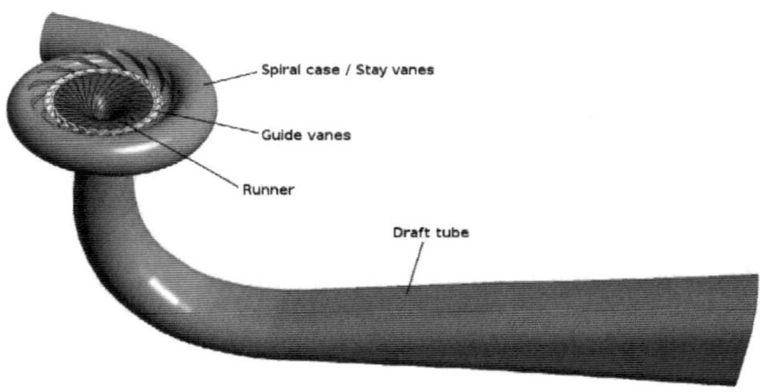

(a) 360°-model.

(b) Reduced model.

Figure 1: Domain discretization

guide vane angles are analyzed. The guide vane opening angles are 9.84, 6.72 and 12.43 degrees at BEP, PL and HL, respectively. For further geometrical details of the model see [10]. The mesh size for the complete model is approx. 23 million elements. Except for the spiral case/stay vane domain, all parts have a structured mesh containing hexahedra elements. Information concerning mesh size and quality criteria can be found in table 1. All values are related to the model at BEP.

Table 1: Mesh parameters for the complete 360°-model for the BEP.

Mesh	Element size	$Y+_{AV}$ (BEP)	min. Angle	Volume change	Aspect ratio
Global	22'984'188	42.2	2.2°	1.90	1.82
Spiral & stay vanes	3'427'708*	30.8	2.2°	1.90	1.82
Guide vanes	7'884'800	42.2	25°	1.20	0.25
Runner	10'780'800	36.2	36.9°	1.23	0.19
Draft tube	890'880	35.5	36.7°	1.18	0.31

Table 2: Element types in the spiral case/stay vane domain*.

Tetrahedra	Pyramids	Wedges	Hexahedra
118'354	154'995	920'880	2'233'479

The definition of the quality parameters is according to [2]. The reduced model for the stationary-case contains a quarter of the complete 360°-model elements. It has approx. 5 million elements. The rotor side spaces are neglected in both models so that disc friction as well as leakage flow are not directly available in the simulations. To compare the simulation with the measurements, the given experimental data for the friction torque from the test case are subtracted from the simulation output.

2.5. Numerical schemes

All the convective terms are discretized with a high resolution scheme. Therefore the flow solver itself switches between first and second order depending on the local flow-parameters [8]. The turbulence schemes are of first order and the temporal term is of second order (implicit Euler backward). To reach sufficient convergence 10 non-linear inner iterations are applied for the unsteady simulation, resulting in rms values for the flow solver residuals of at least 10^{-5}. Initially the time step was chosen in a way, so that the temporal resolution of simulation was equal to 3° of a rotor revolution. The value was then further reduced until the highest frequencies of the pressure signal in the simulation could be captured (approx. 10 samples per period were required). A sufficient sampling of the pressure signal had been achieved with a time step 0.75ms, which corresponds to a resolution of 1.5° per each time step.

2.6. Initial and boundary conditions

For the steady state simulation no solution from a previous CFD simulation is used for the initialization. Therefore the simulation started from zero using constant values for k, ω, U and p in the initialization. All unsteady simulations are initialized with a steady state Multiple Reference Frames (MRF) solution.

At the inlet of the computational domain the mass flow rate according to the operating point, a turbulent kinetic energy based on the turbulent intensity I of 5%, which is a common value for hydraulic turbines [6] and a turbulent frequency based on $\left(\frac{\mu_t}{\mu}\right)$ around 10 for confined flow conditions [3] are imposed.

At the outlet of the computational domain, an average static pressure of $p_{\text{outlet}} = 0Pa$ is applied, due to the incompressible flow type. The walls in the domains are non-rotating, except for the runner domain. At the walls, a no-slip condition is imposed. The runner domain rotates with constant speed around the z-axis.

3. Results

3.1. Integral values

3.1.1. Comparison between measurement and simulation

In order to evaluate the accuracy of the simulation, the two simulation-models with constant boundary condition are analyzed and compared with the steady measurements. In figures 2 (a)-(c) the values for the efficiency, the torque on the runner and the net head are shown. The measured torque represents the torque on the shaft. In consideration of the measurement set-up the following definition for the net head was used:

$$H = \left(\left(\frac{1}{A_{in}}\right)^2 - \left(\frac{1}{A_{out}}\right)^2\right) \cdot \frac{Q^2}{2g} + (p_{AV,in} - p_{AV,out}) \cdot \frac{1}{\rho g} \tag{3}$$

IOP Conf. Series: Journal of Physics: Conf. Series **782** (2017) 012007 doi:10.1088/1742-6596/782/1/012007

H stands for the net head and A_{in} as well as A_{out} represent the cross sectional areas on the inlet respectively outlet of the turbine. Further $p_{AV,in}$ and $p_{AV,out}$ are the area averages of the static pressure on the inlet, respectively outlet, Q represents the volume flow rate across the turbine and g stands for the gravitational constant. It can be seen that the values for the

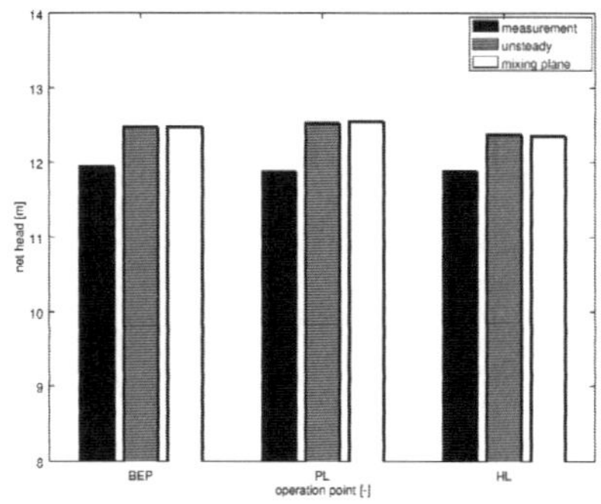

(a) Net head.

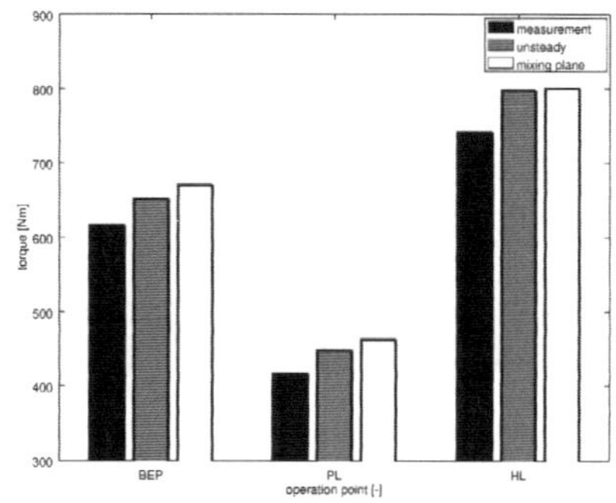

(b) Torque on the runner.

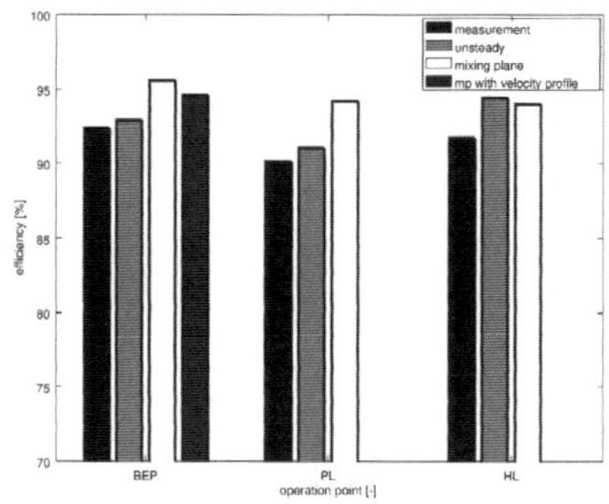

(c) Overall efficiency.

Figure 2: Comparison of the global variables between measurement and simulation.

head in all simulations have been overestimated compared to the measurements. A possible explanation for this behavior is, that at the inlet of the turbine, a volume flow rate with a constant velocity distribution is imposed. The inlet pipe is relatively short (see fig.1) and therefore the flow field does not have sufficient time to develop a fully turbulent velocity profile before entering the bladed region of the machine. Therefore, a steady state simulation of the BEP is conducted using a simulated fully developed velocity profile. The result in figure 2 (c) shows that a difference of 1% can be observed in the efficiency between profile boundary condition and constant value boundary condition, which brings simulation results and measurements closer together. Therefore it is presumed, that a difference can also be expected for the unsteady simulations. The difference in efficiency results from 1% less torque and 0.16% less net head using the fully developed velocity profile. However, the relative difference of 0.16% in head does not justify the gap between measurement and simulation. Further reasons which are not examined here such as mesh sensitivity and turbulence model influence would have to be analyzed in for closure.

For the torque it can be seen, that the stationary simulation with the mixing plane overestimates the value by a significant amount. The unsteady simulation is closer to the measurements except for the high-load operating point.

The resulting values for the efficiency of the unsteady simulation correspond well with the measurement. The error made for the BEP is just 0.61%. Even for PL the error is relatively small with 0.90%. Under high load conditions the simulation model has more difficulty to capture the flow behavior. For this operating point is the error for the efficiency 2.74%. The steady state simulations consequently shows larger differences to the measurement. The errors are 3.16%, 4.12% and 2.31% for BEP, PL and HL, respectively.

3.1.2. Comparison between the steady state and unsteady simulations As outlined above, there is a significant discrepancy between the computed torque from the steady state simulation and the unsteady simulation for BEP and PL conditions. The reason can be traced back to the simplification made for the steady state model. An unsteady operation point is simulated as if it behaves steady. With the mixing plane approach mentioned above, the flow parameters are averaged along the circumference at the interfaces. For homogeneous flows, like for example for a turbine at the BEP, the influence is usually small [4]. Nevertheless in the simulation it can be stated, that even for the BEP, the flow field contains areas of flow separations at the stay vane trailing edge, which causes the differences in the torque. See fig.3. The separation

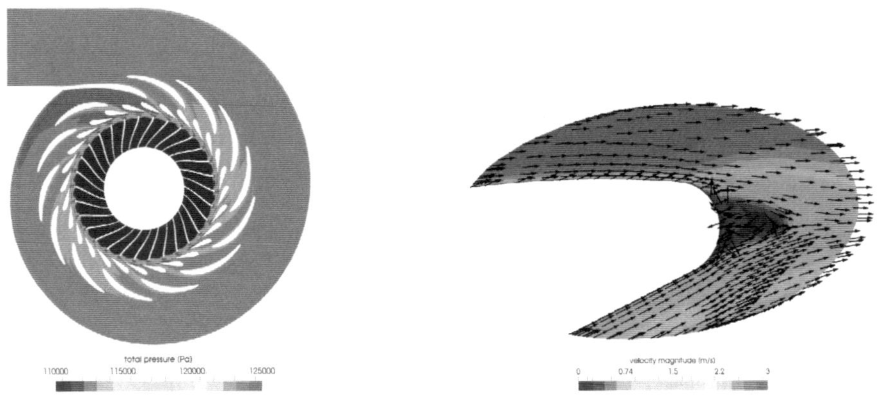

(a) Development of the absolute total pressure on the mid-height (z=0m).

(b) vector-and conturplot of the absolute velocity at the stay vane trailing edge.

Figure 3: Detected separation zone for the unsteady simulation at BEP

zone influences the flow field and leads to a marginally different incidence angle at the runner. The relative difference in flowangle of the absolute velocities at the interface between runner and guide vanes is 2.6% between the two models at BEP. The resulting pressure profile on the runner blades consequently leads to a different value for the torque relative to the undisturbed flow field. The averaging process of the mixing plane reduces the intensity of this effect.

3.2. Pressure probes

Time-dependent pressure measurements are provided for this study, from two sensors placed in the draft tube short after the trailing edge of the runner and a third one in the vaneless space (fig. 4 (a)). To compare the measurements with the simulations, only the unsteady simulations are used. In order to visualize the dominant frequencies, Fast Fourier Transformations (FFT) have been performed.

In the vaneless space the dominant frequencies mainly correspond to the blade passing frequencies for all operating points. Therefore there are peaks at 15 and 30 times the runner frequency, corresponding to the 15 splitter blades and 15 runner blades. In fig.5 (a) and (b) data for PL is reported. It can be seen, that the simulation results contain the same dominant frequencies like the measurement, but not vice versa. Also, the time-signals show similar behavior. It can also be noted that the measured signals have a larger magnitude and the FFT show a much broader spectrum than the simulation. Part of the broader spectrum is caused by the system itself but also by the measurement noise. Because the simulation model does not consider the measurement set up of the turbine with all its armatures (compare figure 1 with figure 4 (b)), the corresponding frequencies are not represented in the FFT spectrum. Due to this reason and the strong band noise in the signal, in the following only the simulated pressure signals are analyzed.

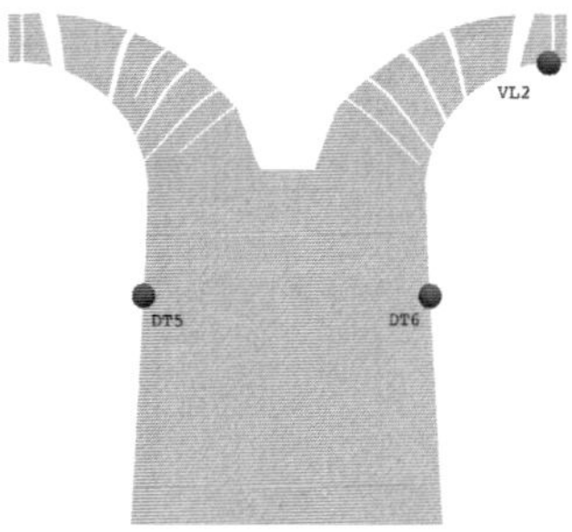

(a) Location of the pressure sensors in the draft tube and vaneless space.

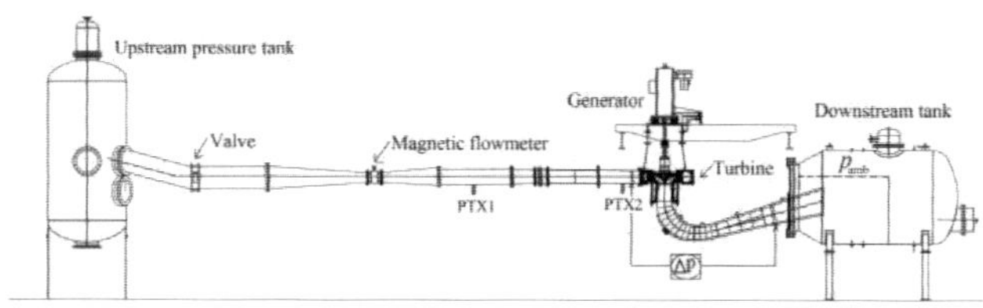

(b) Model Francis turbine test rig [11].

Figure 4: Measurement set-up.

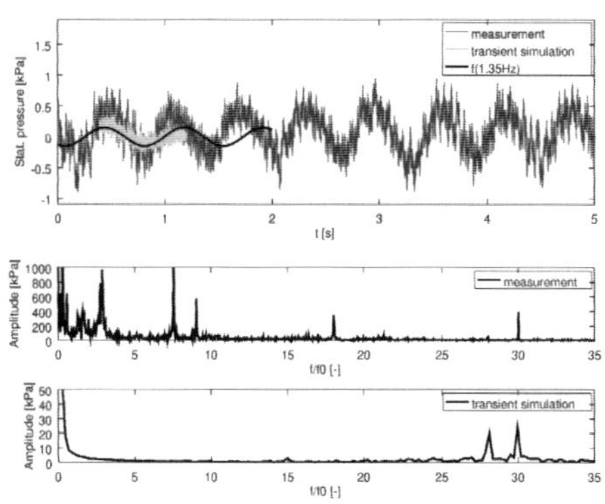

(a) Pressure signal in the draft tube (DT5).

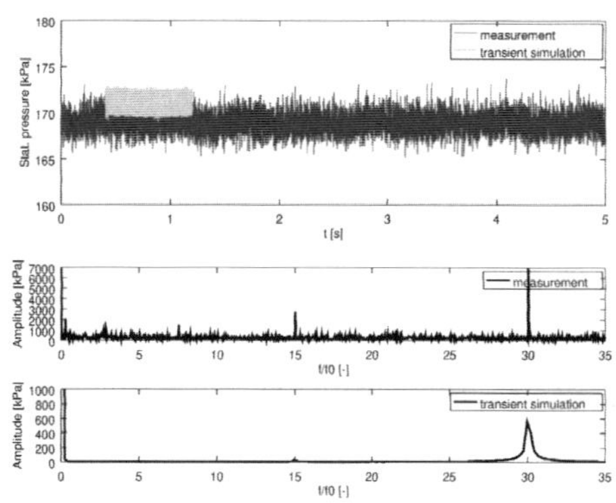

(b) Pressure signal in the vaneless space (VL2).

Figure 5: Comparison of the pressure signal (time signal and FFT) between the measurement and the unsteady simulation for PL.

In the following analysis of the pressure signal the normalized pressure was used to obtain a dimensionless representation for the spectrum. Therefore the vertical axis represents the normalized pressure amplitude, which is defined as $\frac{p-p_{\text{average}}}{p_{\text{average}}}$. On the horizontal axis the frequency divided by the rotational frequency is used.

3.2.1. BEP and HL

For BEP and HL the dominant frequencies at the sensor position in the vaneless space are represented by the blade passing frequency, fig.6 (a)&(b). Contrary to the pressure signal in the vaneless space the frequencies in the draft tube are less dominant and therefore had to be recorded over a longer period of time, in order to get enough information to distinguish between the separate peaks.

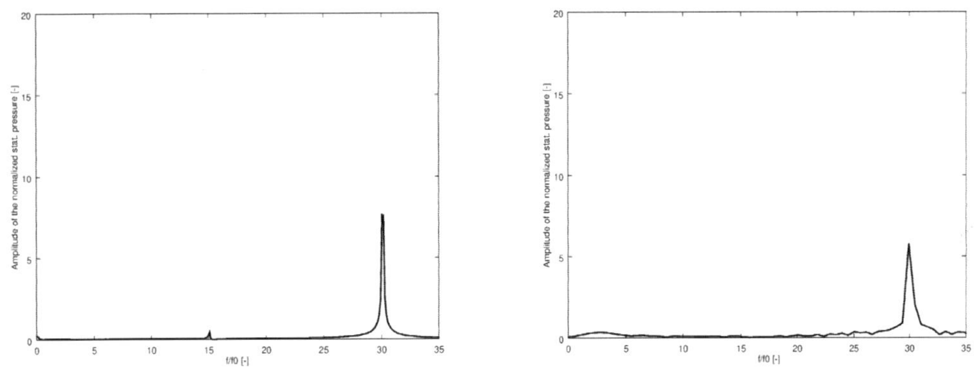

(a) BEP. (b) HL.

Figure 6: Pressure signal in the vaneless space for BEP and HL (VL2).

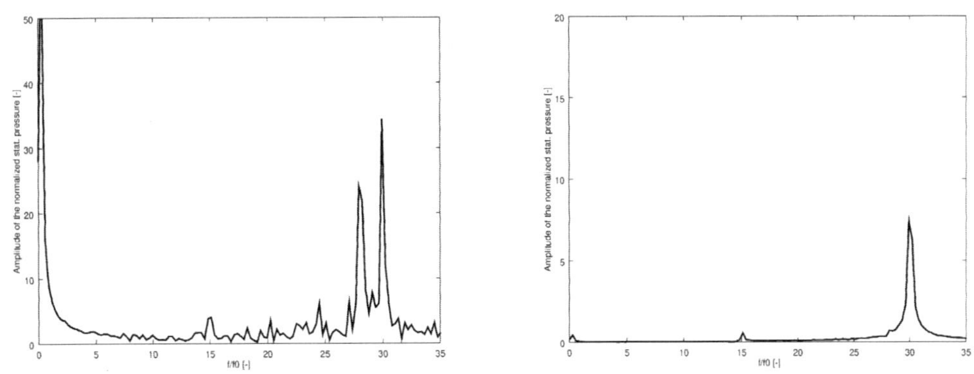

(a) FFT-analysis in the draft tube (DT5). (b) FFT-analysis in the vaneless space (VL2).

Figure 7: FFT-analysis of the pressure signals at PL condition.

3.2.2. PL

The dominant frequencies for PL condition also correspond to the blade passing frequencies in the vaneless space, fig.7 (b). Even in the draft tube location these frequencies can be detected (fig.7 (a)). For the draft tube pressure signal there is also a remarkable peak at 155Hz, which corresponds to 28 times the rotation frequency. For an accurate identification of the origin of this frequency it is necessary to record additional data, which requires additional investigations.

Again, also for the pressure signal at PL condition the recording duration was relatively small, fig.5 (a)&(b). This is one reason why not all significant frequencies can be captured and represented by the FFT. In the time signal of fig.5 (a) another distinct oscillation can be recognized for the measurement as well as for the simulation. In the figure this oscillation is represented by the black solid line. This relatively slow oscillation of the simulated pressure signal has a frequency of approx. 1.2-1.4Hz which correlates to 20%-25% of the runner frequency. The frequency can be attributed to the vortex rope in the draft tube cone [1].

3.3. Velocity probes

In the second Francis 99 workshop, velocity measurements in the draft tube are provided. The measurements were realized with the use of Particle Image Velocimetry (PIV). Unlike in the first workshop, the velocity-components in axial and radial direction are given, instead of the tangential component. The measurement is given on three sampled lines, two in the horizontal direction and one in the vertical direction at the center of the draft tube cone, fig.8. For more detailed information about the measurement set up see [10]. Because of the small amplitudes of the radial velocity component only the axial velocity is analyzed in the following. Subsequently

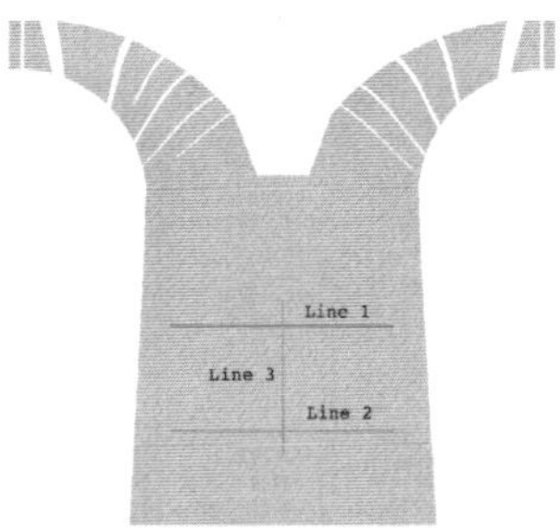

(a) Measurement set-up on the meridional section.

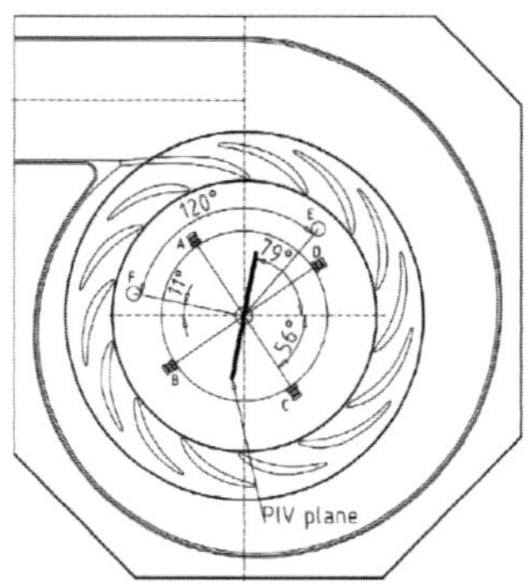

(b) PIV plane from topview [10]

Figure 8: Location of the measuring lines in the draft tube cone.

the averaged velocity component in axial direction is plotted over the sampled lines. Because the measurements are recorded with a relatively low sampling rate of 40Hz, the complete provided data set was used for the averaging over time. This corresponds to 200 discrete values for each point on the line. An evaluation of the signal for just one channel rotation ($\frac{360^\circ}{30}$) could not be performed, because there are only seven samples per runner rotation and therefore the signal can not be captured properly. For this reason the simulated velocity signals were averaged over one runner revolution for each point on the line. For each point 240 discrete values are used for the averaging. The velocity components are normalized with the average meridian velocity (volume flow rate divided by the area at the draft tube entry). The horizontal values are plotted from left to right and the vertical ones from bottom to top.

3.3.1. BEP

Line 1&2 (horizontal):

For the BEP it can be shown, that the axial flow behavior can be captured with the unsteady and the steady state simulation. One reason for the good agreement of the steady state simulation with the measurements is caused by the axial symmetric flow behavior in the draft tube which is given at the BEP. For this case the averaging over the circumference of the draft tube entry leads to no additional error. An other reason for the good agreement between simulation and measurement is due to the relatively small velocity gradients along the measured line (fig. 9) at BEP conditions. Therefore it is easier for the simulation model to capture the flow field here, rather than in other operation points.

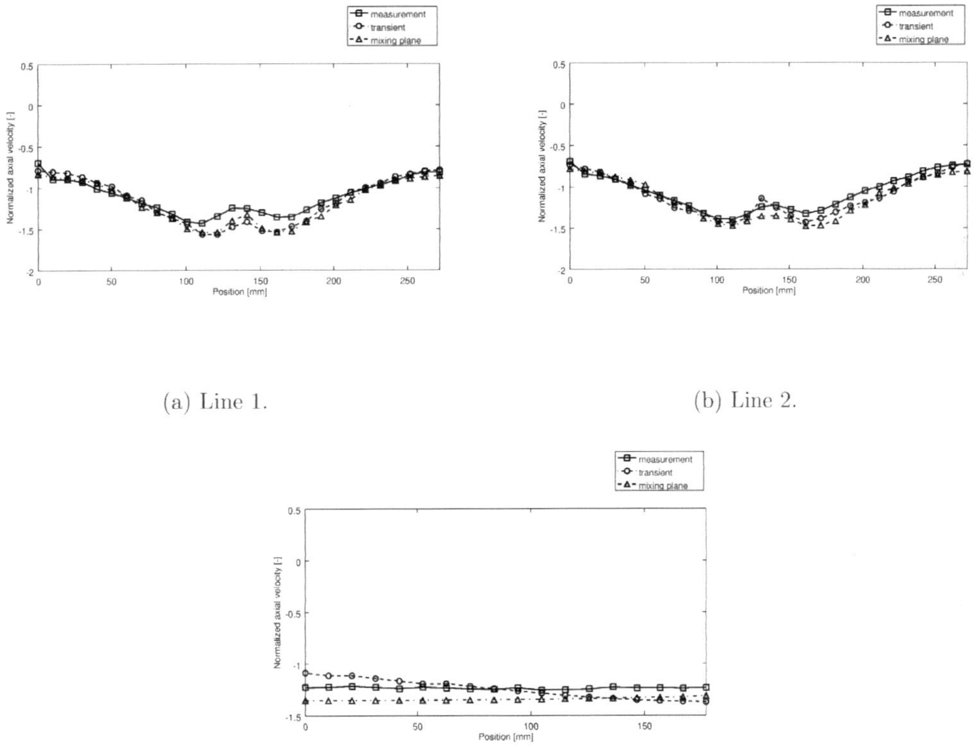

(a) Line 1.

(b) Line 2.

(c) Line 3.

Figure 9: Normalized axial velocity at BEP.

Line 3 (vertical):
The measured values on the vertical line have no characteristic profile. It can be seen that for the axial velocity component there is an offset between simulation and experiment. Qualitatively, the plots are comparable.

3.3.2. HL

Line 1&2 (horizontal):
Compared with the BEP-case the operation point at high load condition shows much higher velocity gradients across the measured line, fig.10 (a). This leads to greater discrepancies between

measurement and simulation. Both, the unsteady and the steady state approach, have more problems to capture the flow field. Nevertheless both are capable to qualitatively reproduce the profile of the flow and also generate a solution on which the physical behavior is comparable to that of the measurement. Again the simulation model with mixing plane can benefit from axial symmetric flow behavior like the BEP.

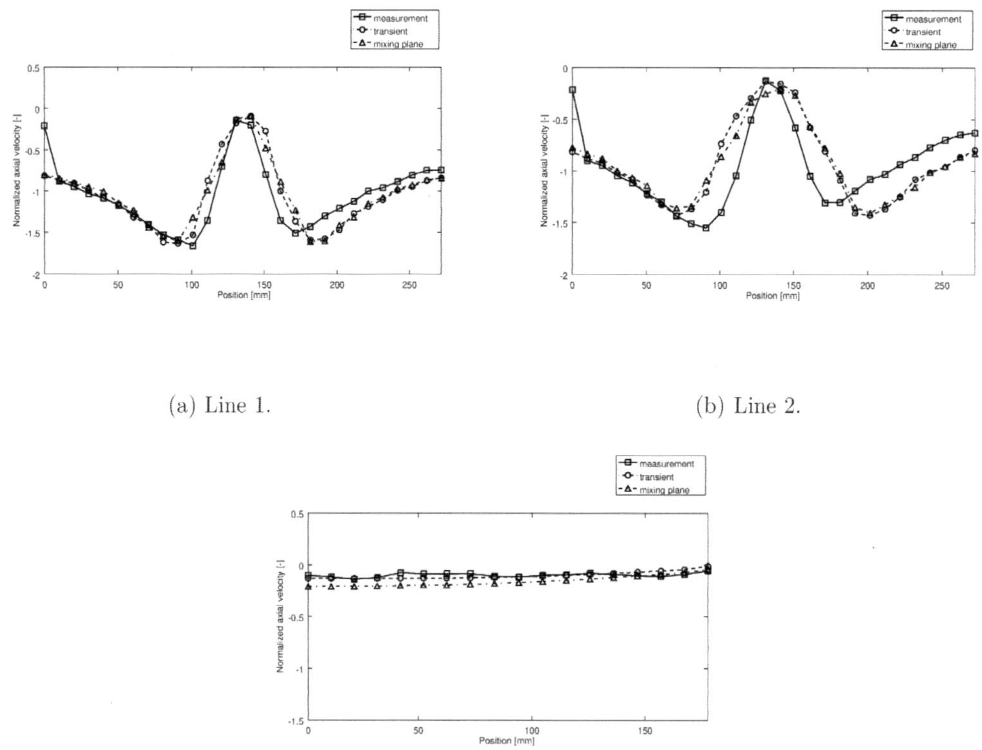

(a) Line 1.

(b) Line 2.

(c) Line 3.

Figure 10: Normalized axial velocity at HL.

Line 1&2 (horizontal):
For PL condition a different flow behavior between measurement and simulation can be recognized. The simulation shows a backflow caused by the vortex rope in the draft tube. For both steady state and unsteady simulations the vortex rope is much more distinct than in the measurement. Remarkable is also, that the steady state simulation is capable to predict the draft tube flow with a similar accuracy as the unsteady simulation. The reason is, as already observed in the last workshop [1], that the vortex rope in the upper part of the draft tube is nearly axisymmetric. Therefore, averaging over the circumference again has a much smaller influence compared to a flow field with an excentric vortex rope attached to the runner.

Line 3 (vertical):

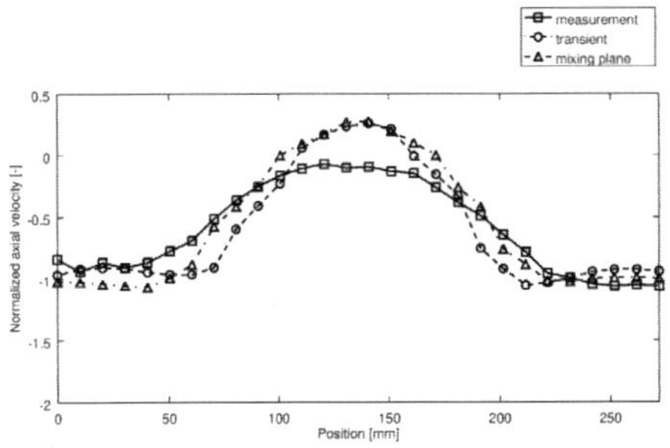

(a) Line 1.

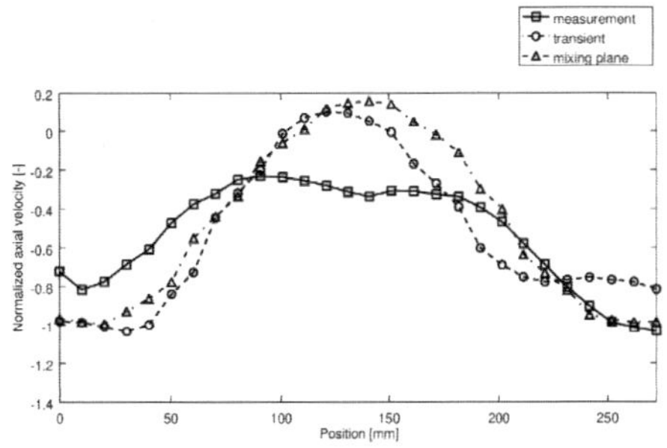

(b) Line 2.

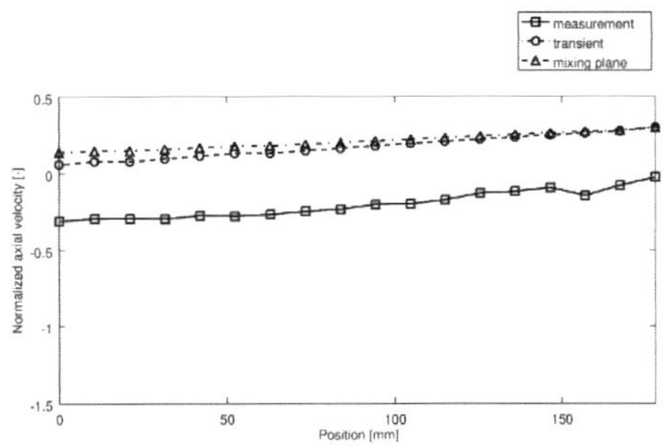

(c) Line 3.

Figure 11: Normalized axial velocity at PL.

The backflow region, which can only be observed for the simulation results, relates to an offset and opposite sign for the axial velocity component compared to the measurement for the vertical line.

3.3.3. Vortex rope

In fig. 13 it can be seen, that for the locations of the velocity measurements the vortex rope for PL condition is almost in the center of the draft tube cone. The vortex has been found by applying the an isosurface with constant value for the Q-criterion on the simulation results. The Q-criterion is defined as positive second invariant for $\nabla \mathbf{u}$. It represents the local balance between shear strain rate and vorticity magnitude. Vortices are areas where the vorticity magnitude is greater than the magnitude of rate-of-strain [5].

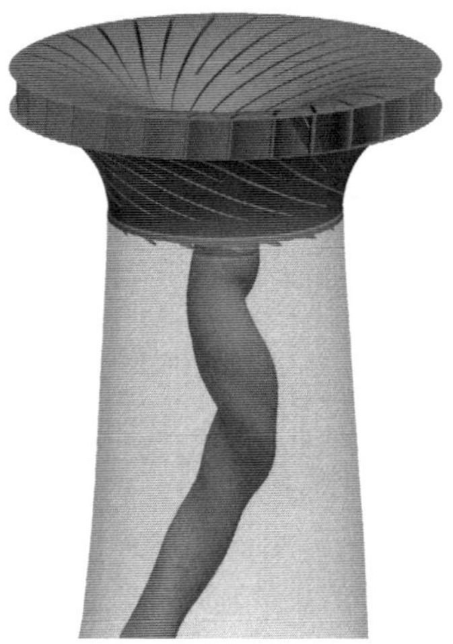

Figure 12: Vortex rope at PL condition visualized with surface of constant value (100) for the Q-criterion.

Conclusion

The CFD simulations could successfully be conducted within the scope of the Francis 99 workshop. Steady-state case and the simulation with the moving runner both overestimate the integral quantities of the baseline measurements provided for this study in terms of torque and head. Since head increases the hydraulic power as much as higher torque increases the mechanical power, the hydraulic efficiency is only marginally different between simulations and measurements. However, unlike the head which is virtually the same for steady state vs. unsteady, the torque is dissimilar between steady state and unsteady throughout the study. A possible reason is found in the different rotor-stator interface treatment. The mixing plane at the steady state interface interferes with separation zones at the stay vanes upstream of the interface. This results in higher torque output for the steady state simulations, especially for PL and BEP.

Furthermore, FFT evaluations of the pressure signals from the simulations and from the measurements have shown that a large amount of frequencies are not reproducible with CFD due to effects caused by the setup of the measurement rig itself. As a result, only few frequencies can be compared among CFD and measurement. Mainly the blade passing frequency for BEP,PL and HL at 30 times the runner frequency and the frequency of the vortex rope in PL at 0.2 times the runner frequency can be detected with certainty. An additional frequency at approx. 28 times the runner frequency in the draft tube at PL can not be associated with any known physical effects and still has to be investigated in detail.

In addition, time-averaged values for axial velocity are evaluated here. The difference between the operating points can clearly be seen when considering the axial velocity on the three sampling lines within the draft tube. The axial velocity component on the axis of rotation confirms the presence of the vortex rope which is already indicated by the FFT evaluation of the pressure probes.

In conclusion, the different effects found here can be observed with both, steady state and unsteady simulations, which is especially noteworthy for the vortex rope due to its axial-centric properties for this PL operating point here. However, a small penalty in error of torque due to the averaging process in the mixing plane of the steady state simulation is inevitable. Overall, BEP and HL agree best with the velocity measurements whereas the PL simulations are more critical in accuracy due to the increased complexity of the flow field. In essence, the steady state simulations of the three operating points, using a coupled incompressible solver, provide accurate information with manageable use of computational resources.

References

[1] B Aakti, O Amstutz, E Casartelli, G Romanelli, and L Mangani. On the performance of a high head francis turbine at design and off-design conditions. In *Journal of Physics: Conference Series*, volume 579, page 012010. IOP Publishing, 2015.

[2] Ansys. *ANSYS ICEM CFD Help Manual*. -, Canonsburg, - edition, 2013.

[3] Users Guide FLUENT. Fluent inc. *Lebanon, USA*, 2003.

[4] L Hanimann, L Mangani, E Casartelli, T Mokulys, and S Mauri. Development of a novel mixing plane interface using a fully implicit averaging for stage analysis. *Journal of Turbomachinery*, 136(8):081010, 2014.

[5] V Kolář. Vortex identification: New requirements and limitations. *International journal of heat and fluid flow*, 28(4):638–652, 2007.

[6] MV Magnoli. Numerical simulation of pressure oscillations in francis turbine runners. *JASS Numerical Simulation of Turbomachinary*, pages 32–50, 2009.

[7] L Mangani, M Buchmayr, and M Darwish. Development of a novel fully coupled solver in openfoam: Steady-state incompressible turbulent flows. *Numerical Heat Transfer, Part B: Fundamentals*, 66(1):1–20, 2014.

[8] F Moukalled, L Mangani, and M Darwish. *The Finite Volume Method in Computational Fluid Dynamics: An Advanced Introduction with OpenFOAM® and Matlab*, volume 113. Springer, 2015.

[9] NTNU and LTU. Francis-99 workshop. *http://www.ltu.se/research/subjects/Stromningslara/Konferenser/Francis-99?l=en accessed august 2014*.

[10] NVKS. Francis-99 second workshop. *https://www.ntnu.edu/nvks/second-workshop accessed september 2016*.

[11] C Trivedi, MJ Cervantes, BK Gandhi, and GO Dahlhaug. Experimental investigations of transient pressure variations in a high head model francis turbine during start-up and shutdown. *Journal of Hydrodynamics, Ser. B*, 26(2):277–290, 2014.

[12] JP Yan, U Seidel, and J Koutnik. Numerical simulation of hydrodynamics in a pump-turbine at off-design operating conditions in turbine mode. In *IOP Conference Series: Earth and Environmental Science*, volume 15, page 032041. IOP Publishing, 2012.

Efficiency limit factor analysis for the Francis-99 hydraulic turbine

Y Zeng[1], L X Zhang[2], J P Guo[3], Y K Guo[4], Q L Pan[5], J Qian[6]

1. Faculty of Metallurgical and Energy Engineering, Kunming University of Science and Technology, Kunming 650093, China. Email: zengyun001@163.com
2. Department of Engineering Mechanics, Kunming University of Science and Technology, Kunming 650500, China.Email: zlxzcc@126.com
3. Yunnan Institute of Water Resources and Hydropower Engineering Investigation, Kunming 650053, Yunnan, China, Email: okgjp@126.com
4. School of Engineering, University of Bradford, BD7 1DP, UK. Email: y.guo16@bradford.ac.uk
5. Sinohydro Bureau 14 Co. LTD, Kunming 650500, China. Email: 13888359265@163.com
6. College of Electric Power Engineering, Kunming University of Science and Technology, Kunming 650500, China.Email: qj0117@163.com

Abstract: The energy loss in hydraulic turbine is the most direct factor that affects the efficiency of the hydraulic turbine. Based on the analysis theory of inner energy loss of hydraulic turbine, combining the measurement data of the Francis-99, this paper calculates characteristic parameters of inner energy loss of the hydraulic turbine, and establishes the calculation model of the hydraulic turbine power. Taken the start-up test conditions given by Francis-99 as case, characteristics of the inner energy of the hydraulic turbine in transient and transformation law are researched. Further, analyzing mechanical friction in hydraulic turbine, we think that main ingredients of mechanical friction loss is the rotation friction loss between rotating runner and water body, and defined as the inner mechanical friction loss. The calculation method of the inner mechanical friction loss is given roughly. Our purpose is that explore and research the method and way increasing transformation efficiency of water flow by means of analysis energy losses in hydraulic turbine.

Keywords: inner energy loss; power model; Francis-99; mechanical frication; efficiency

List of symbols

c_1 ,c_2	undetermined coefficient	Q_z	flow discharge at the peak efficiency point in m^3s^{-1}
C_f	viscosity resistance coefficient	R_e	Reynolds number

$F(Q-Q_z)$	impact loss function	Q_{BEF}	Flow discharge at peak efficiency point in m^3s^{-1}
f_p	coefficient of pipe friction loss	R_f	friction resistance in N
H_0	hydrostatic head of hydraulic turbine in m	S	wet area in m^2
H_b	Basic value of the head in m	T_{fr}	Friction torque in Nm
H_t	head of the hydro turbine in m	T_t	Torque to generator in Nm
k_h	coefficient of flow channel loss	α	Guide vane angle (°)
k_v	coefficient of volume loss	γ	water density in $kg\cdot m^{-3}$
M_f	rotation friction torque in Nm	Δh_q	is the transient head change in transient process in relative
n	Rotation speed in rpm	ω	Angular speed in rad/s
n_b	Basic value of angular speed in rpm	ΔP_h	iturbine loss power in the flow channel in kW
P_t	output power of the hydraulic turbine in kW	ΔP_i	impact loss power in kW
P_{tz}	turbine power at the maximum efficiency point.	ΔP_m	mechanical friction loss power in kW
P_{t0}	Basic value of power in kW	ΔP_v	volume loss power in kW
Q	turbine flow discharge in m^3s^{-1}	η_1, η_2	efficiency of hydro turbine
Q_b	Basic value of flow discharge in m^3s^{-1}	μ	dynamic viscosity of water in $N\cdot s\cdot m^{-2}$

1. Introductions

It is an effective path that research energy loss characteristics of hydraulic turbine, then take steps to decrease loss and improve transfer efficiency of water energy. Loss characteristics of hydraulic turbine is one of research hotspot.

Usually, the inner energy loss of hydraulic turbine is divided into the volume loss, the hydraulic loss and the mechanical friction loss. The physical conception volume loss is the most clear. However, effect of the volume loss on efficiency of hydraulic turbine is usually ignored due to its value too small. The mechanical friction loss includes two parts, one is the external mechanical friction loss produced by friction resistance between bearing and shafting of hydraulic turbine generating unit, another is the internal mechanical friction loss generated by rotating resistance, which is the resistance between water body and rotating runner of turbine in [1,2]. Although calculation external mechanical friction is difficult, it could be obtained by test and measurement, such as the external mechanical friction torque is provided in Francis-99 in [3]. However, the friction resistance between rotating runner and water body is called as the internal mechanical friction loss, its action mechanism is different from the disk loss of centrifugal pump, and its calculation and test method is not reported in the literature. Taken the hydraulic loss as core, calculation and test research in [4,5] are the most active research field. Especially, with the development of computational fluid dynamics (CFD), all aspects of hydraulic turbine characteristics, such as numerical simulation of the hydraulic turbine passageway in

[6,7,8], optimization design in [9,10,11], performance prediction in [12,13,14], and character change of runner blade in [15,16,17], have obtained many achivement. On account of the limitation of computing resource and inner flow regime complexity of turbine runner, the CFD using to research transient process of hydraulic turbine still has some limitations in [18].

A calculation method of inner energy loss of hydraulic turbine is proposed in Ref.[19], which can be extracted loss characteristics of hydraulic turbine from test data, and it has been obtained new extensive application in [20]. Based on this method, using data provided by the Francis-99 test platform, the energy loss characteristics of hydraulic turbine are investigated. Meanwhile, calculation method of the internal mechanical friction loss is proposed, although it is an attempt and exploration.

2. Power calculation model of hydraulic turbine

2.1 Basic model

Refer to the turbine model with the inner energy loss characteristics in [19], the inner energy of the hydro turbine composition can be written as following:

$$P_t = \gamma Q H_t - \Delta P_h - \Delta P_v - \Delta P_m - \Delta P_i \tag{1}$$

Where γQH_t is the water flow power of hydraulic turbine, $\Delta P_\gamma = \gamma QH_t k_v$ is the volume loss power, the$\Delta P_h = k_h Q^3$ is turbine loss power in the flow channel.

At the optimum operation of the hydro turbine, there is no water flow impacting on the inlet of the turbine runner while water outflow moves along the normal direction at the outlet of the turbine runner. Under non-optimum operation, there exists the impact loss at both the inlet and outlet of the turbine runner. From the efficiency curve of the hydro turbine, the impact loss is almost the same for the same deviation of the flow rate away from the peak efficiency point. Therefore, it is possible to combine the impact loss at the inlet and the outlet, the effect of the draft tube and other losses except clear losses in Eq.(1) into the impact loss power ΔP_i. As such, ΔP_i is a function of flow and can be expressed as $\Delta P_i = F[(Q-Q_z)]$.

The power expression of hydraulic turbine can be written as follow:

$$P_t = \gamma H_t Q(1-k_v) - k_h Q^3 - \Delta P_m - F[(Q-Q_z)] \tag{2}$$

Remark 1: The head of hydraulic turbine H_t in Eq.(2) can be calculated by using the form given by the IEEE Working Group in [22], that is $H_t = H_0(1-f_p q^2 - \Delta h_q)$, $f_p q^2$ is the friction loss head in pipeline in relative, in steady state, $\Delta h_q = 0$. Thus the Eq.(2) is suitable for both the large and small fluctuations under arbitrary head.

2.2 Computation of characteristic parameters

Using the data under steady state operation to calculate the characteristic parameters. Let $h_q=0$; the coefficient of volume loss k_v can be selected according to the seal characteristics of the turbine. Usually, k_v is taken as 0.0025~ 0.005.

(1) Determination of parameter k_h

Based on the definition of the impact loss power in Remark 1, assume that the impact loss be zero at the peak efficiency point for the same head and the impact loss be the same for the same flow deviation away from the peak efficiency point. Selecting the efficiency curve at some water head and taking a point at each side of the maximum efficiency point, the flow discharge satisfies the following.

$$\begin{cases} Q_1 = Q_z + \Delta Q \\ Q_2 = Q_z - \Delta Q \end{cases} \tag{3}$$

The impact loss at these two points is equal to each other. From Eq. (2), one can obtain:

$$k_h = \frac{\gamma H_t (Q_1 - Q_2)(1-k_v) - P_{t1} + P_{t2}}{Q_1^3 - Q_2^3} \quad (4)$$

where $P_{t1}=9.81Q_1H\eta_1$, $P_{t2}=9.81Q_2H\eta_2$, η_1 and η_2 is the efficiency corresponding to flow Q_1, and Q_2, respectively.

(2) Mechanical friction loss

Assume the impact loss ΔP_i=0 at the maximum efficiency point, the loss power due to the mechanical friction at the maximum efficiency point can be calculated as:

$$\Delta P_m = \gamma H Q_z (1-k_v) - k_h Q_z^3 - P_{tz} \quad (5)$$

(3) Impact loss

The impact loss power can be written as:

$$\Delta P_i = \gamma Q H_t - \Delta P_m - \Delta P_h - \Delta P_v - P_t \quad (6)$$

Given flow discharge Q, the impact loss power can be calculated by combining Eq.(6) with interpolation calculation of P-Q curve. Taking $(Q-Q_z)$ as variable, the impact loss power is further fitted to impact loss function $F(Q-Q_z)$.

3. Francis-99 example

Whole data of four operation points provided by the Francis-99 test case are list as table 1.

Table 1. Steady state measurements data

Parameter	PL	BEP	HL	Min load
Guide vane angle α (°)	6.72	9.84	12.43	0.80
Net head H_t (m)	11.87	11.94	11.88	12.14
Discharge Q (m^3s^{-1})	0.140	0.200	0.242	0.022
Torque to generator T_t (Nm)	426.39	616.13	740.54	11.16
Friction torque T_{fr} (Nm)	4.40	4.52	3.85	4.66
Runner rotation speed n (rpm)	332.84	332.59	332.59	332.80
Hydraulic efficiency η (%)	90.13	92.39	91.71	20.94

The g is the gravity, g=9.882($m \cdot s^{-2}$), γ=999.8($kg \cdot m^{-3}$) is the water density, γ=9.88($kN \cdot m^{-3}$). The PL is the operation point of part load, the BEP is the operation point of the peak efficiency point, and the HL is the operation point of high load.

Using data of four operation points, power model and power loss of hydraulic turbine will be calculated in next sections.

Remark 2. The classical calculation equation of hydraulic turbine is $P_t=\gamma QH\eta$(kW), γ=9.81($kN \cdot m^{-3}$). In the Francis-99 test case, the γ is 9.88($kN \cdot m^{-3}$). If power of hydraulic turbine is calculated by the equation $P_t=9.88QH\eta$(kW), while test power is calculated by the equation $P_t=n\pi T_t/30000$ (kW), maximum power error at the peak efficiency point is 1.58%。Analysis revealed, the maximum power error at the peak efficiency point is 0.86%, where it uses the γ=9.81($kN \cdot m^{-3}$). Therefore, we use γ=9.81($kN \cdot m^{-3}$) to calculate in follow-up sections.

Hydraulic turbine power changing with flow discharge is shown as Figure 1.

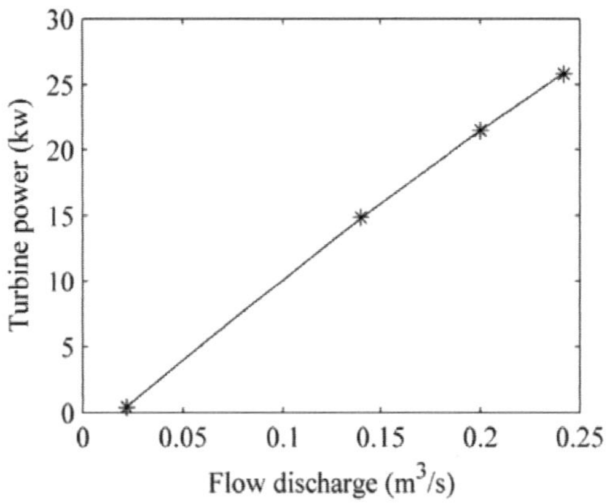

Figure 1. Turbine power changing with flow discharge

In figure 1, the sign * denotes test data, while the solid line is fitting curve.

Efficiency curve can be obtained by test data at PL, BEP, HL and Min load. In order to decrease error, one uses test data at PL, BEP and HL to obtain the fitting curve of efficiency top parts. Using the top curve calculation of interpolation to obtain new data, and combining with data at Min load to compose new data group, and then the efficiency curve can be obtained by fitting curve. The efficiency curve is shown as figure 2.

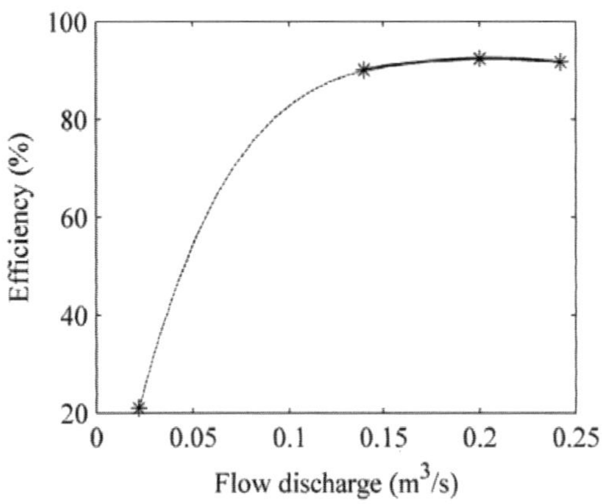

Figure 2. Efficiency curve of the hydraulic turbine

In figure 2, the sign * denotes test data, while the solid line is fitting curve.

Remark 3: The purpose of forming efficiency curve is using the classical equation $P_t=\gamma QH\eta$ to calculate two side power at the peak efficiency point, and then the coefficient k_h can be calculated by the Eq.(4). However, there are only test data of four operation point in the Francis-99 test case. Obviously, using test data of four point to fit the efficiency curve has certain subjectivity, and maybe exist large error. For this reason, the efficiency curve is not applied to use in later calculation in this paper. Form the Eq.(4), the calculation of coefficient k_h only requires to two side power at the peak efficiency point. From figure 1, the power of hydraulic turbine changing with the flow discharge is near to straight line. The turbine power expression can be obtained by combining polynomial of degree 1 and degree 2. So, the power fitting curve as figure 1 is used to calculate power of hydraulic turbine near to the peak efficiency point, while calculates the coefficient k_h.

4 Francis-99 model of power calculation

In order to analysis relative change of parameters, according to data given by Ref.[5] determine parameters basic value system as, H_b=12.00 m, Q_b=0.20 m^3s^{-1}, P_{t0}=22.00 kW, ω_b=333 rpm. In fact, purpose of selection basic value system is that reflect relative change of calculation parameters, and change is more perceptual intuition.

The volume loss coefficient of hydro turbine k_v is near selected as zero, that is k_v=0.

In order to improve the computational accuracy, ten unit flow increments are selected as ΔQ=±0.001, 0.002, 0.003, 0.005, 0.01 m^3s^{-1}. The corresponding coefficient k_h is calculated using the ten increments. The mean value is k_h=60.4641.

Assumption the impact loss power at the peak efficiency point is zero, the mechanical friction loss power can be obtained from Eq.(5), ΔP_m=1.4823kW, or 6.74%P_{t0}.

According to the Q-P curve shown as figure 1, and given flow discharge Q, the impact loss power can be extracted by Eq.(6). The impact loss power changing with flow discharge is shown as figure 3.

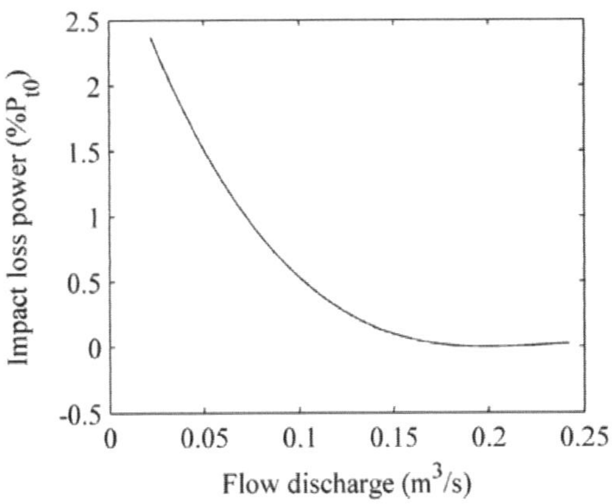

Figure 3. Impact loss power changing with flow discharge

Taking (Q-Q_z) as variable, the impact loss power can be further fitted to the impact loss function $F(Q-Q_z)$.

$$F(Q-Q_{BEF}) = -60.4641(Q-Q_{BEF})^3 + 5.6980(Q-Q_{BEF})^2 + 0.0017(Q-Q_{BEF}) \tag{7}$$

Figure 3 shows that impact loss power is direct relation to degree of deviation peak efficiency point. Especially, the proportion of impact loss in low load regions is large.

Thus, the power expression of hydraulic turbine Eq.(2) can be rewritten as following.

$$P_t = 9.81H_tQ - 60.4641Q^3 - 1.4823 - F(Q-Q_{BEF}) \tag{8}$$

5. Energy loss characteristics of hydraulic turbine

Taken data the Min Load to PL as case to calculation, its sample rate is 5 kS/s, but its data fluctuation is large. To this reason, after analyzing various digital filtering method, this paper uses the polynomial fitting of degree 30 to smoothness sample data. The head of hydraulic turbine change in transient is shown as figure 4(a), and the flow discharge in transient is shown as figure 4 (b).

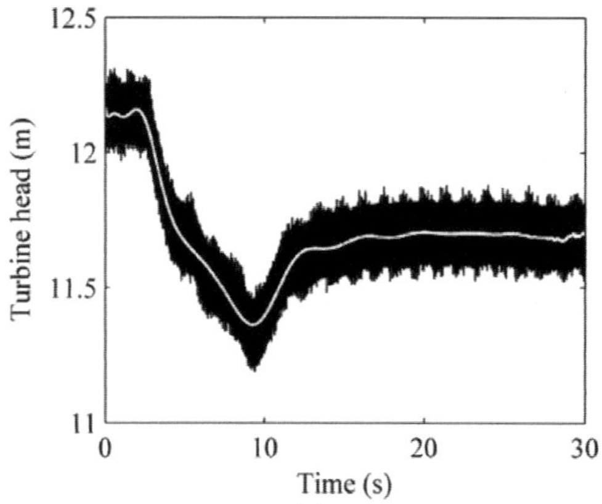

(a)

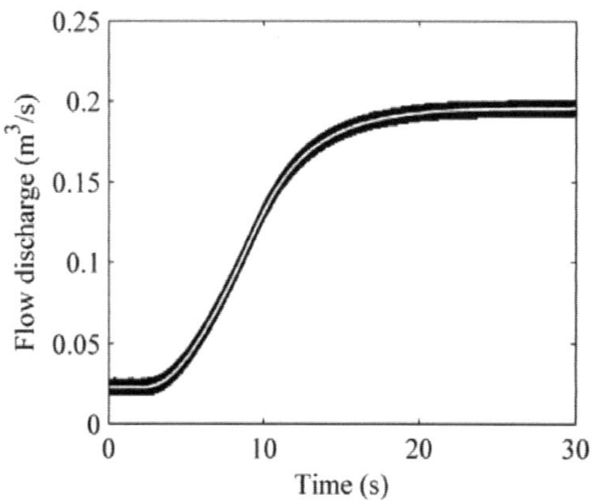

(b)

Figure 4 Sampling data pretreatment

In figure 4, solid line is smoothing data, which shows that using this kind of higher degree polynomial fitting method to smooth sample data is feasible.

Using test data of head and flow discharge in transient, the power of hydraulic turbine can be calculated by Eq.(8), it is shown as Figure 5.

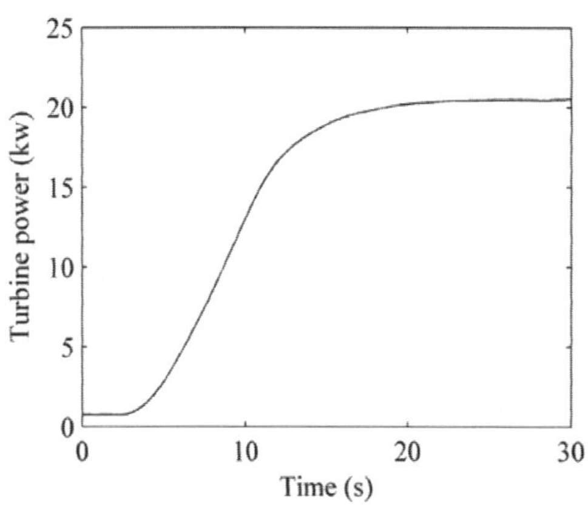

Figure 5 Turbine power changes in transient

Taken the P_{t0}=22 kW as basic value, the energy loss change in startup process is shown as Figure 6.

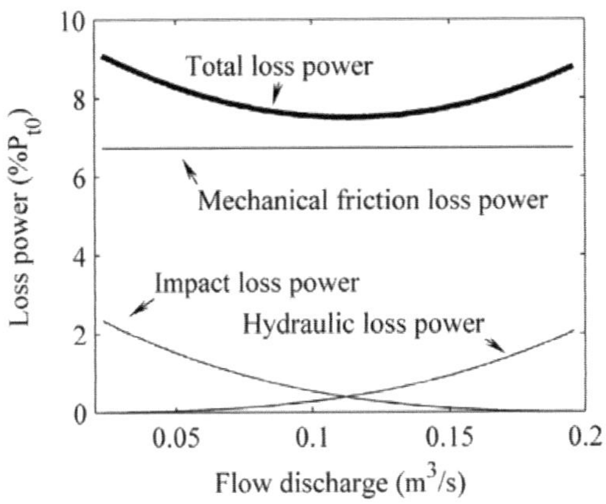

Figure 6. Characteristics of power loss

Remark 4. In other calculation example, the minimum value of total loss power is near the peak efficiency point, In Francis-99 case, the flow discharge in the peak efficiency point is 0.2m^3/s, where the minimum value of total loss is not near it. The reason maybe include two parts, (1) amplitude of head variation in test is small, thus its impact loss is low; (2) The hydraulic loss defined in model (1) is the hydraulic loss of runner passageway loss. The runner diameter of Francis-99 is 349 mm, its size is small. With flow discharge increase, the hydraulic loss in runner regional sharply increase, its maximum is about 2.06%P_{t0} , means the hydraulic loss is leaning large.

Remark 5. The loss energy of runner passageway increase with flow discharge, and it relate with the three power of flow discharge. Therefore, by means of optimization design of runner passageway and elaborate processing to decrease loss is an effective measure, especially in high load zone.

6 Mechanical friction loss analysis

6.1 Loss composion

The mechanical friction loss power defined in Eq.(1) is direct related to rotational speed of hydraulic turbine generating units. Under units connecting to power network, fluctuation amplitude of rotated speed of units is very small. As a result, the mechanical friction loss power in transient is approximately invariant, it is about 6.74%P_{t0}. It accounts for a large proportion, and remains invariant within the scope of the whole power change .

There are three kinds of mechanical friction in runner zone of hydraulic turbine.

(1) The friction torque in different operation point in table 1 are 4.40, 4.52, 3.85 and 4.66 Nm respectively. Transforming them into power and using relative value form, they are 0.70, 0.72, 0.61, 0.74 %P_{t0} respectively.

Obviously, friction torque are obtain by measuring thrust bearing and guide bearing in Francis-99 test, which is the friction torque between rotation system of units and support system of bearing, that is external mechanical friction torque.

(2) One of link generating power loss is the sealing ring of runner. Its calculation power loss is complex, thus it is not calculated in this paper. Total loss in sealing ring is about 0.5%~0.8%P_{t0} refer to Ref.[21], so it is approximately selected as 0.8% P_{t0} in our calculation.

(3) Internal mechanical friction loss. The internal mechanical friction loss is that the friction between rotating runner and water body. In this case, total mechanical friction loss power is 6.74%P_{t0},

so the internal mechanical friction loss power can be indirectly obtained, that is $5.24\%P_{t0}$=$6.74\%P_{t0}$−$0.70\%P_{t0}$−$0.8\%P_{t0}$.

From view of loss numerical value, the internal mechanical friction loss is main ingredients of total loss, it need to be analyzed in the furture.

6.2 Internal mechanical friction loss

The wheel as a whole, the rotation friction surface of wheel rotating part and water body include four parts , inlet edge of blade, outlet edge of blade, runner cone and the parts region of bottom ring, as Figure 7.

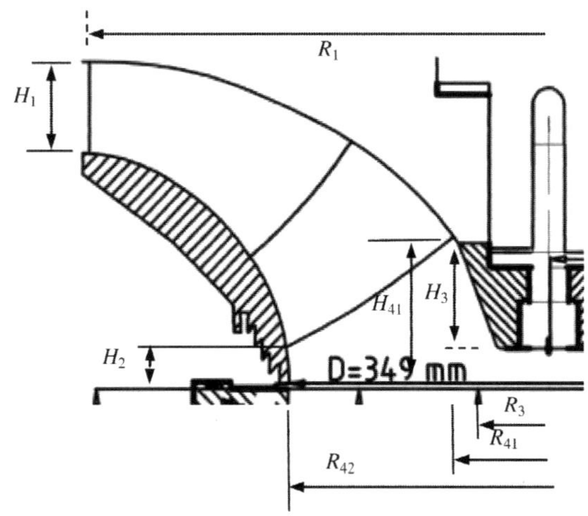

Figure 7. Schematic diagram of rotation friction surface

Due to edge of rotation surface is irregular shape, rotation friction force should be calculated by selecting infinitesimal element and integrating method. Instant motion of infinitesimal element is similar to motion form of fluid flowing through the stationary flat. Therefore, refer to Prandtl-Schlichting equation, the friction resistance coefficient of rotation motion can be approximately expressed as following.

$$C_f = \frac{c_1}{(\lg \mathrm{Re})^{c_2}} \tag{9}$$

$$R_e = \frac{\gamma r^2 \omega}{\mu} \tag{10}$$

where r is the distance from infinitesimal element to axes.

$$R_f = \frac{1}{2} C_f \gamma v^2 S \tag{11}$$

where v is the speed, the dynamic viscosity in water temperature 20° C is μ=1.005×10^{-3} N·s·m^{-2}.

Remark 6: The Eq.(9) is obtained by refer to the Prandtl-Schlichting equation, where coefficient c_1 and c_2 is undetermined coefficient. Currently, there are not systematic test analysis and reference for the rotation resistance of hydraulic turbine. According to the empirical formula in planar turbulent boundary layer, let c_1=0.455 and c_2=2.58. We think that rotation motion of turbine runner is similar to motion of rotor disc, so calculation in this paper uses c_1=0.455 and c_2=2.58.

Friction resistance of any infinitesimal element in rotation edge can be expressed as following.

$$\mathrm{d}R_f = \frac{1}{2} C_f \gamma v^2 \mathrm{d}S = \frac{1}{2} C_f \gamma \omega^2 r^2 \mathrm{d}S \tag{12}$$

And friction torque is

$$M_f = \iint_S r\mathrm{d}R_f = \frac{1}{2}\iint_S C_f \gamma\omega^2 r^3 \mathrm{d}S \tag{13}$$

In inlet water edge, $r=R_1$, the wet aera of infinitesimal element is $\mathrm{d}S_1=R_1\mathrm{d}\theta_1\mathrm{d}h_1$. θ_1 is the angle of infinitesimal element motion, h_1 is the height of infinitesimal element. Assume that the Reynolds number in inlet water edge is constant, the resistance torque in inlet water edge of runner can be obtained by integral operation.

$$M_{f1} = \iint_S r\mathrm{d}R_{f1} = \frac{1}{2}C_f \gamma R_1^4 \omega^2 \iint_S \mathrm{d}\theta_1 \mathrm{d}h_1 = C_f \pi\gamma\omega^2 H_1 R_1^4 \tag{14}$$

Lower edge in outlet water of the bottom ring and runner cone can be approximately account as rotation cylinder. Their friction torque can be calculated using the Eq.(14).

At radius r of outlet water edge, the wet aera of infinitesimal element is $\mathrm{d}S_4=r\mathrm{d}\theta_4\mathrm{d}h_4$. Shape of outlet water edge is considered as straight line approximately. Its rotation friction torque is.

$$M_{f4} = \iint_S r\mathrm{d}R_f = \frac{1}{2}C_f\gamma\omega^2 \iint_S r^4 \mathrm{d}\theta_4 \mathrm{d}h_4 = \frac{1}{5}\pi C_f \gamma\omega^2 \frac{h_{42}}{R_{42}-R_{41}}(R_{42}^5 - R_{41}^5) \tag{15}$$

Total rotation friction torque can be written as following.

$$\begin{aligned} M_f &= M_{f1} + M_{f2} + M_{f3} + M_{f4} \\ &= C_f \pi\gamma\omega^2 \left[H_1R_1^4 + H_2R_2^4 + H_3R_3^4 + \frac{1}{5}\frac{h_{42}}{R_{42}-R_{41}}(R_{42}^5 - R_{41}^5) \right] \end{aligned} \tag{16}$$

Remark 7. The Eq.(16) shows that internal mechanical friction loss of the Francis turbine is closely related to geometric structure of runner and blade. This feature provides important basis for decreasing internal mechanical friction loss and increase efficiency of hydraulic turbine.

Some of structure sizes can be obtained from runner draw of Francis-99, H_1=0.057m, R_1=0.315m, H_2=0.020m, R_2=0.175m, H_3=0.070m, R_3=0.050m, H_{42}=0.095m, R_{42}=0.175m, H_{41}=0.165m, R_{41}=0.065m. Ignoring influence of splitter blades, the equivalent radius of outlet water edge of runner is $(R_{42}^2-R_{41}^2)^{0.5}$=0.1625m.

Rotation resistance torque of each parts can be calculated by the Eq.(14) and Eq.(15), and are listed as table 2.

Table 2. Rotation resistance torque of each parts

Rotary surface	1	2	3	4
Radius (m)	0.3150	0.1750	0.0500	0.1625
Reynolds number ($\times 10^6$)	3.4422	1.0624	0.0867	0.9161
C_f ($\times 10^{-3}$)	3.5839	4.4206	7.3895	4.5448
M_{fi} (Nm)	7.6822	0.3167	0.0123	0.7458
Ratio of resistance power (%Pt0)	1.2177	0.0502	0.0020	0.1182

Remark 8. Total resistance power by summing losses of four parts is 1.3881%P_{t0}. It is less than the 5.24%P_{t0} obtained in section 5.2. However, we find that 4×1.3881%P_{t0}=5.5524%P_{t0}, it is close to the 5.24%P_{t0} obtained in section 5.2. Why is this so?. Althought we don't still give reasonable excuse, but we think that it is related to the calculation method of the coefficient C_f. For example, the inlet water edge is accounted as a whole surface of rotation, and it's the coefficient C_f is calculated by means of the pannel viscosity resistance method. So calculation of the coefficient C_f should be improved in the furture.

Table 2 reveals composition ration of the internal mechanical friction loss power in the Francis-99, which provides more intuitional calculation basis and reference for optimization design based on geometric structure of turbine runner.

7. Conclusions

By subdividing internal energy loss of hydraulic turbine and taking Francis-99 as case, detailed calculation and extraction method of energy loss of hydraulic turbine are given. Using algebra model form of hydraulic turbine power research changes of energy loss characteristic in transient, that provides a theoretical basis for analysis affect factors and limitation conditions of hydraulic turbine efficiency.

According to analysis examples of Francis-99, two points conclusions can be given as following.

(1) Using proposed method decomposing internal energy loss of hydraulic turbine and investigating its changes in transient process, can more clear reveal ratio of losses and their transformation law, which provides more direct reference for optimization design of hydraulic turbine.

(2) Rotation friction loss of hydraulic turbine has larger ratio in energy loss, and relates to geometric structure closely, that is worth attention. Derivation of rotation friction loss given by this paper is just a attempt, in which certain parameters selection and approximate treatment not yet sufficient theoretical foundation, they need further research.

Acknowledgments

The research reported here is financially supported by the National Natural Science Foundation of China under Grant No. 51579124, 51469011,51279071. The constructive and insightful comments and suggestions made by two anonymous Reviewers have significantly improved the quality of the final manuscript.

References

[1] Cheng Liangjun. Hydraulic Turbine[M]. Beijing, China Machine Press, 1981.
[2] Liu Dakai. Hydraulic Turbine[M]. Nanjing, China WaterPower Press, 1980.
[3] External mechanical friction torque, in table 3, http://www.ntnu.edu/nvks/test-case
[4] Qian Hanxin, Liu Shuhong, Lin Ruchang. Energy conversion characteristics of mixed flow water turbines with high specific speed[J]. Power Engineering, 1996, 16(5), 56-60,49,44
[5] Chirag Trivedi, Michel J. Cervantes, and Ole G. Dahlhaug. Experimental and Numerical Studies of a High-Head Francis Turbine: A Review of the Francis-99 Test Case[J]. Energies, 2016, 9(2), p. 74. http://dx.doi.org/10.3390/en9020074.
[6] Wang Zhengwei, Chen Liu, Xiao Ruofu, Zhang Huanneng, Shao Baoan. Study on influence of the construction change of the spiral casing and draft tube on the flow loss in low head Francis turbine[J]. Journal of Hydrorlrctric Engineering, 2008, 27(5), 147-152,134.
[7] S. Bahrami, C. Tribes, C. Devalsb, T.C. Vuc, F. Guibault. Multi-fidelity shape optimization of hydraulic turbine runner blades using a multi-objective mesh adaptive direct search algorithm[J]. Applied Mathematical Modelling, 2016, 40(2), 1650–1668.
[8] D. Benzon, A. Židonis, A. Panagiotopoulos, G. A. Aggidis, J. S. Anagnostopoulos and D. E. Papantonis. Numerical Investigation of the Spear Valve Configuration on the Performance of Pelton and Turgo Turbine Injectors and Runners[J]. Journal of Fluid Engineering, 2015,137(11), 111201-8.
[9] Luo Xingqi, Liang Wuke, Chen Naixiang, Peng Guoyi, Lin Tuchang. A opitmum design method method of hydraulic turbine runner[J]. Journal of Hydraulic Engineering, 1996, 10, 32-38.
[10] Lai Xide. Hydraulic optimization design for hydraulic turbine based on performance estimation[J]. Journal of Hydraodynamics, A ser. 17(6), 656-664.

[11] Ciocan Tiberiu, Susan-Resiga Romeo F., Muntean Sebastian. Modelling and optimization of the velocity profiles at the draft tube inlet of a Francis turbine within an operating range[J]. Journal of Hydraulic Research, 2016, 54(1), 74-89.

[12] Guo Pengcheng, Luo xingqi, Qin yanchun. Numerical performance prediction for a Francis turbine based on computational fluid dynamics stage simulation[J]. Proceedings of The Chinese Society for Electrical Engineering, 2006, 26(17), 132-137.

[13] Alveyro Teran Leonel, Jose Larrahondo Francisco, Aida Rodriguez Sara. Performance improvement of a 500-kW Francis turbine based on CFD[J]. Renewable Energy, 2016, 96, 977-992.

[14] Feng Jianjun, Luo Xingqi, Wu Guangkuan, Zhu Guojun. Influence of clearance flow on efficiency prediction of Francis turbines[J]. Transactions of the Chinese Society of Agricultural Engineering, 2015, 31(5), 53-58.

[15] Zhang Lixiang, Wang Wenquan, Yaoji. Analysis of flow-induced vibration of blades for a Franics hydro turbine[J]. Engineering Mechanics, 2007, 24(8), 143-150.

[16] Qian Zhongdong, Li Wan, Huai W.X. Wu Yulin. The effect of runner cone design on pressure oscillation characteristics in a Francis hydraulic turbine[J]. Journal of Power and Energy, 2012, 226(1), 137-150.

[17] Wang Fujun, Liao Cuilin, Tang Xuelin. Numerical simulation of pressure fluctuations in a large Francis turbine runner[J]. Chinese Journal of Mechanical Engineering, English Edition, 2012, 25(6), 1198-1204.

[18] Chirag Trivedi, Michel J. Cervantes, and Ole G. Dahlhaug Numerical Techniques Applied to Hydraulic Turbines a Perspective Review[J]. Applied Mechanics Review, 2016,88 010802-1-18.

[19] Yun Zeng, Yakun Guo, Lixiang Zhang, Tianmao Xu, Hongkui Dong, Torque model of hydro turbine with inner energy loss characteristics[J]. Science China, series E, 2010, 53(10), 2826–2832.

[20] Jing Qian, Yun Zeng, Yakun Guo, Lixiang Zhang. Reconstruction of the complete characteristics of the hydro turbine based on inner energy loss[J]. Nonlinear Dynamics, 2016, 82(2), 963-974.

[21] Harbin Institute of Large Electrical Machinery, Hydraulic Turbine Design Manual[M]. Beijing, China Machine Press,1976.

[22] IEEE Working Group on Prime Mover and Energy Supply Models for System Dynamic Performance Studies: Hydraulic turbine and turbine control models for system dynamic studies. IEEE Trans. Power Syst. 1992, 7(2), 167-179.

Frozen Rotor and Sliding Mesh Models Applied to the 3D Simulation of the Francis-99 Tokke Turbine with Code_Saturne

N Tonello[1], Y Eude[1], B de Laage de Meux[2], M Ferrand[2]

[1] Renuda, 329-339 Putney Bridge Road, London, SW15 2PG, UK
[2] EDF R&D, MFEE, 6, quai Watier -BP 49, 78401 Chatou, France

nicolas.tonello@renuda.com

Abstract. The steady-state operation of the Francis-99, Tokke turbine [1-3] has been simulated numerically at different loads using the open source, CAD and CFD software, SALOME [4] and *Code_Saturne* [5]. The full 3D mesh of the Tokke turbine provided for the Second Francis-99 Workshop has been adapted and modified to work with the solver. Results are compared for the frozen-rotor and the unsteady, conservative sliding mesh approach over three operating points, showing that good agreement with the experimental data is obtained with both models without having to tune the CFD models for each operating point. Approaches to the simulation of transient operation are also presented with results of work in progress.

1. Introduction

The Francis-99 workshops series is motivated by the need for numerical means of carrying out accurate and realistic computer modelling of high-head Francis turbines, to support the design of high efficiency turbines over the expanding operational range demanded by operators [1, 2]. Due to the size of the turbines, their complexity, with moving parts and vanes, tight tolerances, and complex, three-dimensional flows, such simulations remain challenging in terms of physics modelling, numerical stability and efficiency, computer requirements, and the size and duration of the calculations. To make it possible to validate numerical models and establish recommended practices, define the state of the art, and highlight areas where progress would be required, a set of experimental data obtained on the laboratory scale, Tokke turbine has been made publically available [3]. In addition, workshops have been organized for researchers to present the results of their simulations on the turbine test cases, with a focus on specific topics. As part of the Second Workshop, five different topics have been proposed, ranging from the influence of numerical and physical modelling in steady-state operation to uncertainty quantification.

In this paper, we present the research carried out on the influence of numerical modelling on the steady operation of the Tokke turbines for the three operating points studied experimentally: Part Load (PL), Best Efficiency Point (BEP), and High Load (HL). All the CFD modelling, simulations, and analysis have been performed with open source software. The tools in the SALOME platform [4] were used for pre- and post-processing, and the general, 3D, CFD solver, *Code_Saturne* [5] was used for the CFD modelling and the calculations. The setting up of the models was carried out on multicore, local workstations. Full simulations were carried out on up to 224 compute cores on EDF's *Porthos* cluster. All the simulations have been conducted in 3D, over the full, 360° geometry. Simulations have been carried out using both the frozen rotor approximation and, full unsteady, sliding meshes technique. Results are compared for each case between the two modelling options and with the experimental data supplied for the workshop [3]. A further effort is also in progress to simulate the turbine in transient operation, and a summary of this research to date is also presented.

2. CFD Modelling

The CFD models were prepared based on the mesh files made available for the Second Workshop, and further adapted for *Code_Saturne* v4.0.

2.1. Mesh Preparation

Complete, 360° meshes were built for the simulations using the 'cgns' files available for the spiral casing, distributor, runner and draft tube. A view of the complete mesh is given in figure 1 below.

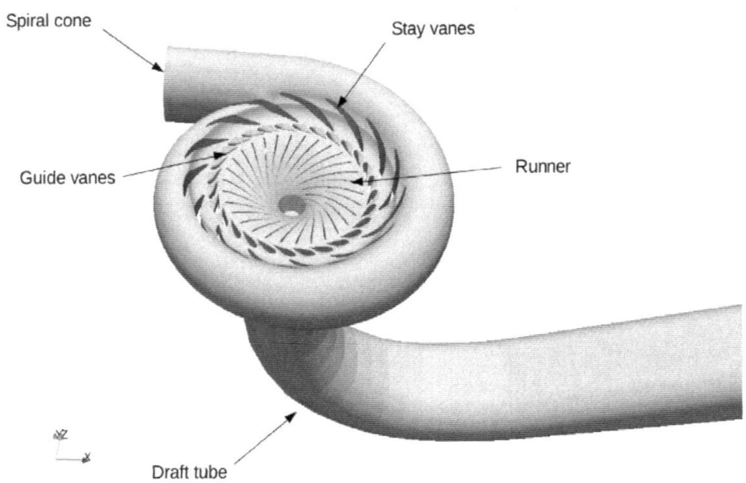

Figure 1: View of the surface of the full computational mesh and the different turbine components.

For the runner, the full wheel was rebuilt based on the single blade and splitting blade passage available in the mesh 'RU-AG5-HighRe.cgns'. The initial mesh was imported in the 'SMESH' module of SALOME and then copied and rotated around a full revolution, with the individual meshes merged into a unique mesh for the wheel. All the operations were carried out using SALOME Python scripts. Further examination of the mesh revealed nodes duplicates in the original mesh. Those were also removed with SALOME, by automatically fusing duplicate nodes (figure 2).

IOP Conf. Series: Journal of Physics: Conf. Series **782** (2017) 012009 doi:10.1088/1742-6596/782/1/012009

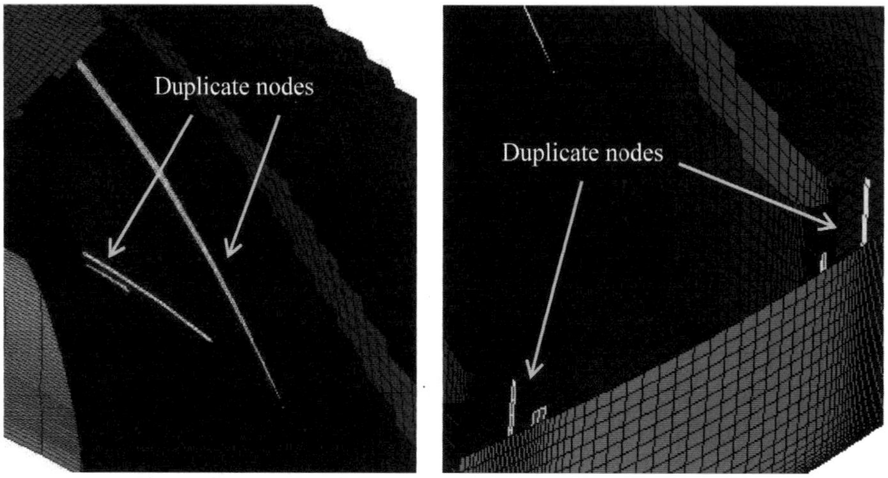

Figure 2: Zoom on duplicate nodes found in the runner mesh.

Having assembled the full 3D mesh, further work was undertaken to build a new interface between the runner and the draft tube. Sliding mesh trials with the initial interface showed that the large difference in discretisation between the runner and the draft tube (figure 3a) at the position of the interface, and the resulting gaps between the two meshes at the inner and outer edges of the interface (figure 3b) were preventing adequate face joining during the mesh sliding operations in *Code_Saturne*.

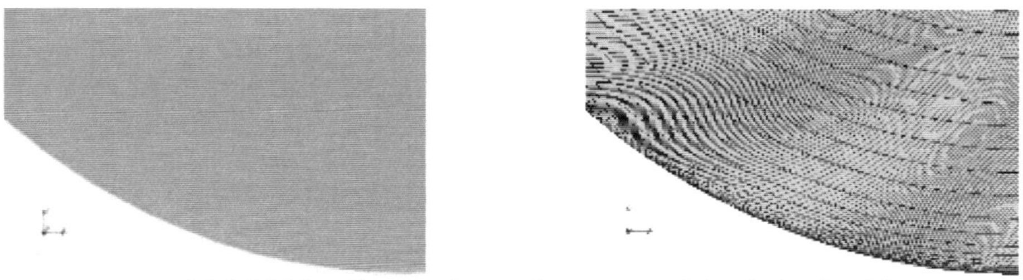

(a) Initial interface meshes on the runner and the draft tube sides

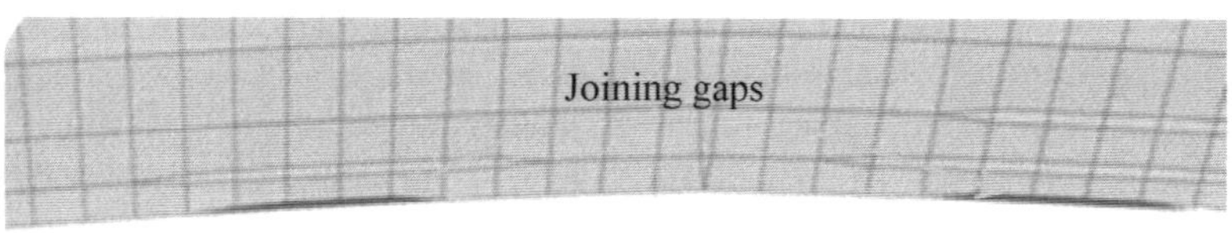

(b) Mesh joining gaps causing errors during the rotation of the sliding interface

Figures 3a-b: Views of the initial meshes at the interface and during rotation.

SALOME scripts were programmed to create a thin slice in the initial mesh, remove the original cells and rebuild a conformal interface, joining the upper side of the new interface mesh to the runner and the bottom side to the draft tube (figures 4a-b).

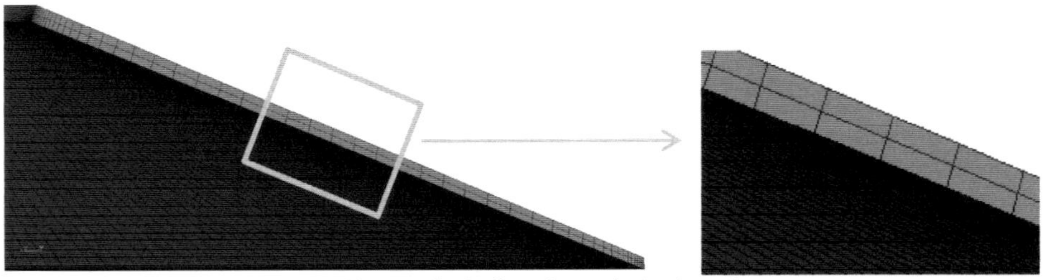

(a) Vertical slice through the interface

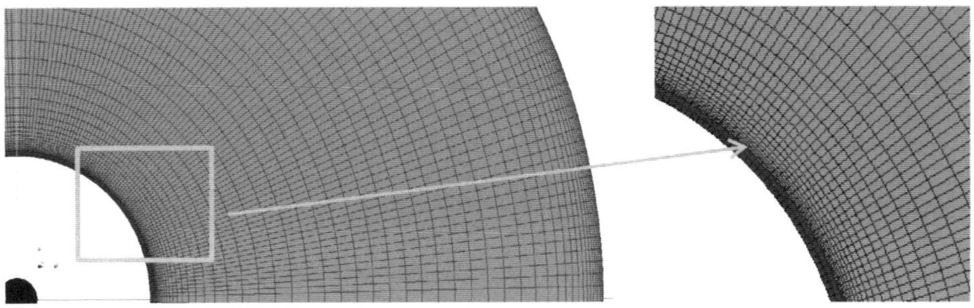

(b) Top view of the interface mesh

Figures 4a-b: Views of the new, two layers, conformal interface mesh.

This final mesh was used for all the calculations, with the frozen rotor and the sliding mesh method.

2.2. CFD Solver Setup

The calculations were setup and run with version 4.0 of the open-source, general CFD solver *Code_Saturne*, developed by EDF R&D. The solver is based on a Finite Volume discretisation on collocated meshes. The Navier-Stokes equations can be solved in three-dimensions, with either RANS or LES approaches for turbulence. A large number of RANS models are available, from algebraic to Reynolds stress or, alternatively, LES can be performed for incompressible or weakly compressible flows. The equations are solved with an implicit, iterative, PISO type algorithm. At the prediction step, the convective and diffusive fluxes are centred, second-order accurate in space. In the correction step, an algebraic, multigrid method is applied to the solution of the pressure equation. The time integration is first-order accurate.

For turbomachine calculations, *Code_Saturne* makes it possible to run both approximated, frozen rotor, and full unsteady, sliding mesh simulations. The frozen rotor simulations are based on the Multiple Reference Frame (MRF) methodology while for sliding meshes approach, the rotation of the rotor mesh is explicitly taken into account.

Two options are available in *Code_Saturne* to account for the rotor mesh rotation. The older methodology consists in coupling two separate instances of *Code_Saturne* running on separate meshes. The interfacial coupling is then handled through specific boundary conditions formulations for the two-instances. The newer method uses a single instance running on the complete mesh, including both static and rotating parts, and uses an advanced mesh-processing algorithm in order to perform the meshes joining. That way, the cell faces on the rotor/stator interface are treated as internal faces, which makes it possible to conserve all the fluxes (mass, momentum, etc.) through the interface. With this sliding mesh method, the rotor mesh coordinates are first updated at the beginning of each time step, and then the mesh joining is performed between the prediction and correction steps of the algorithm, just before the mass flux are updated.

The crux of the conservation properties is the coupled solver which has been implemented in *Code_Saturne*. Rather than solving them one after the other, in the prediction step the momentum equations for all three velocity components are solved together at the same time. This results in one single matrix, which is three times larger than the individual components' equations' matrices but makes it possible to obtain a system where more terms can be written in an implicit, accurate manner. In particular, the boundary conditions at walls and no-slip boundaries can be applied on several components together.

In validation tests, EDF has shown that this new algorithm helped to reduce both solution oscillations and calculation times compared to the original coupling method. Therefore, it is the recommended approach for mesh rotation, provided that the mesh discretisation is similar on each side of the sliding interface in order to allow adequate face joining operations. The algorithm is fully parallelised and has demonstrated linear scaling ability in studies realised by Renuda and EDF.

The sliding mesh simulations discussed in this paper were carried out using this method, which is part of the newly developed turbomachinery module of *Code_Saturne*.

2.3. Numerical and Physical Model

For the single-phase simulations of the Second Workshop, water was modelled as an incompressible fluid with a density of 999.8 kg/m^3 and a viscosity of $9.57 \cdot 10\text{-}7$ m^2/s. The RANS k-ω SST model was applied to simulate the effects of turbulence. Figure 5, taken from the calculations results described in Section 3 below shows that the non-dimensional wall distance y^+ of the first computation node is located either in the viscous sub-layer or in the log-region. Therefore, the k-ω SST model is well-suited to this mesh and it has been preferred to other turbulence models available in *Code_Saturne*.

Figure 5: Non-dimensional wall distance of the first computation node (BEP).

Both frozen rotor and sliding mesh simulations were carried out with an unsteady time marching scheme with a time-step of 10^{-4} s. The maximum CFL in the calculations was approximately 10, with an ensemble average over the cells lower than one and a spatial average of the order of 10^{-2}.

Three seconds of physical time were simulated for frozen rotor computations at each operating points, corresponding to about 16.5 revolutions of the runner. In spite of the absence of runner mesh rotation in the frozen-rotor calculations, some unsteadiness of the flow was observed in the results, especially in the draft tube elbow or downstream of the runner at part load. Thus, a time averaging of the results was performed during the final 4.4 revolutions (0.8 seconds) in order to compute the averaged flow.

The sliding mesh computations were then performed for an additional 1.5 physical second, starting from the corresponding frozen rotor computation, for each operating point. The last 0.8 second was used for the time averaging.

2.4. Computing Resources

The simulations were carried out on the EDF cluster *Porthos* and 224 compute cores. Typically, computations took 24 hours for each second of real time. The calculations times with the sliding mesh modelling were approximately 25% longer than with the frozen rotor approach.

3. Steady-State Operation

Simulations were performed for the three operating conditions (PL, BEP, HL). The numerical results are presented below for the two modelling approaches and compared to the experimental results.

3.1. Velocity

The velocity magnitude computed with the sliding mesh model is shown in figures 6 and 7 below for the BEP and the two off-design conditions, respectively. The flow structure is similar with the frozen rotor model. At BEP, the flow is uniform, without noticeable recirculations. However, at PL and more so for the HL conditions, the vortex tube is unstable and flaps between the walls of the draft tube. At HL, significant non-uniformity and losses can be observed in the spiral cone. Calculating this type of flow on a mixed grid made up of polyhedral and hexahedral cells is challenging for CFD solvers.

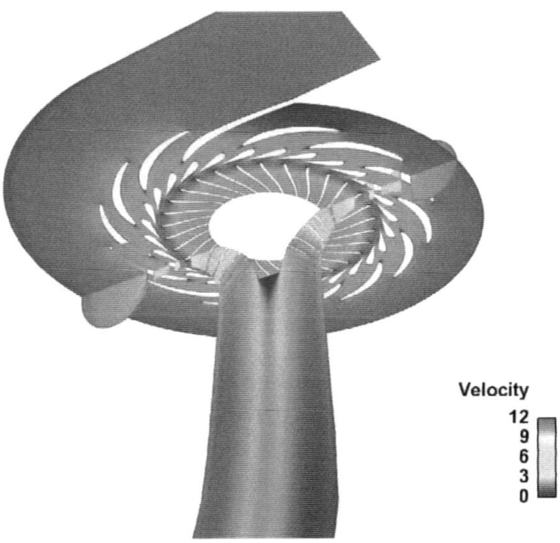

Figure 6: Instantaneous velocity field, sliding mesh simulation. BEP operating condition.

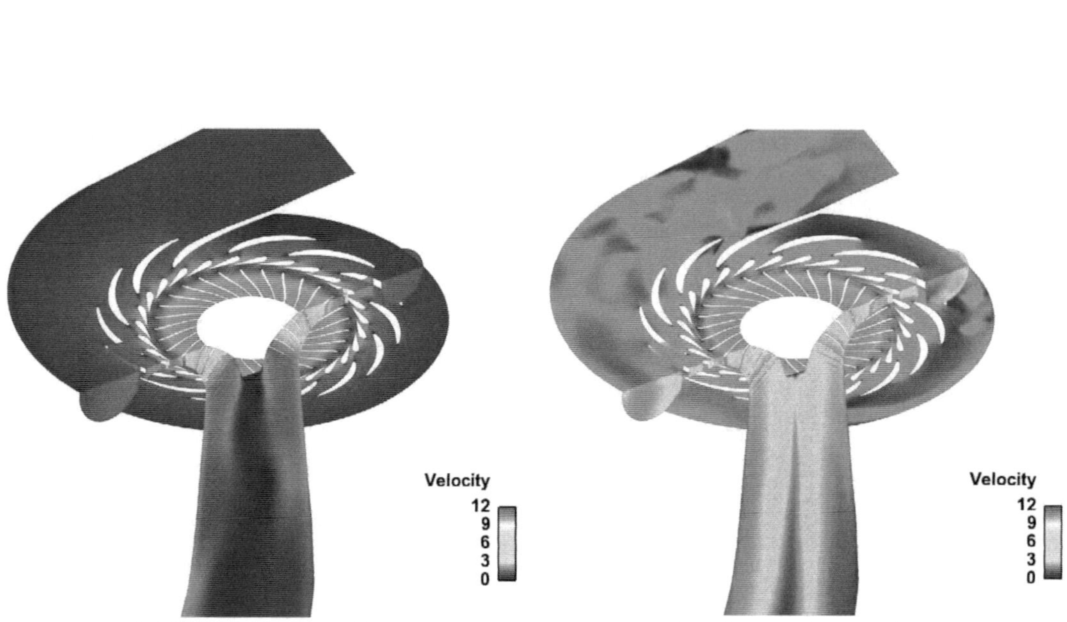

(a) PL operating condition (b) HL operating condition

Figures 7a-b: Instantaneous velocity field, sliding mesh simulation. Off-design conditions.

The numerical results are now compared to the experimental data measured along lines 1 and 2 [3] (figure 8).

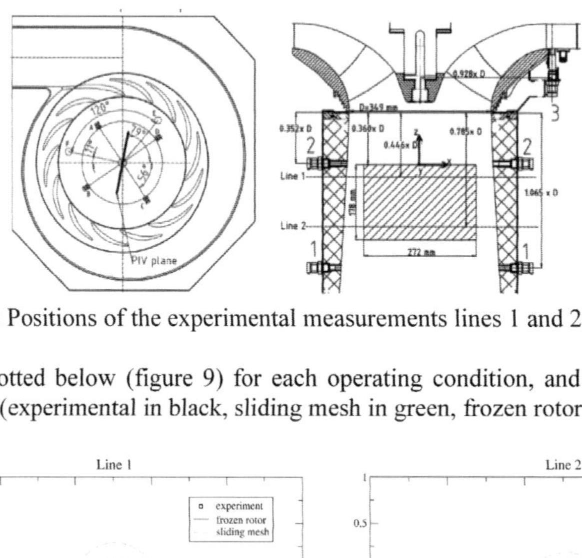

Figure 8: Positions of the experimental measurements lines 1 and 2. From [3].

The results are plotted below (figure 9) for each operating condition, and with the three sets of results on each graph (experimental in black, sliding mesh in green, frozen rotor in red).

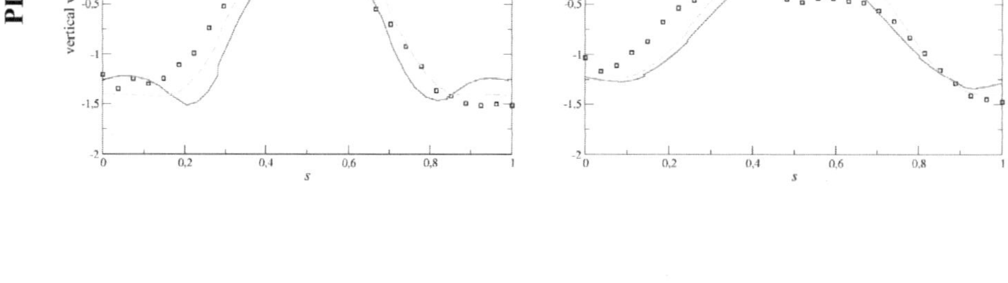

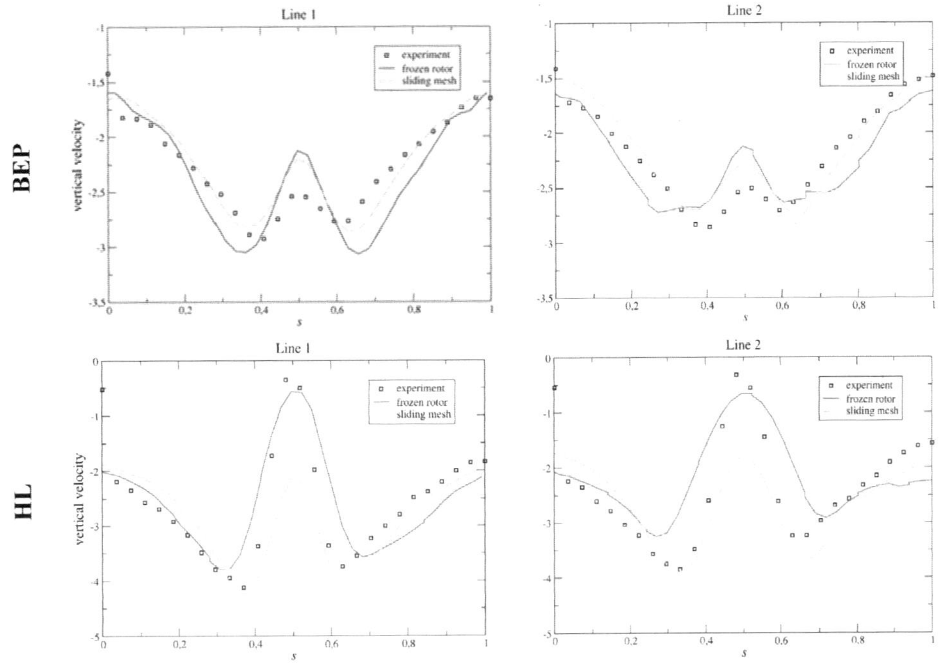

Figure 9: Numerical and experimental axial velocity vs. normalised radius along lines 1 and 2.

Comparing the calculated velocities for the vertical velocity to the experimental measurements shows that both sets of simulations successfully capture the changes in velocity profiles between the three operating conditions and along both measurement lines below the cone. At the Best Efficiency Point, the best results are obtained with the sliding mesh simulations. For all three operating points, the sliding mesh results agree well with the experimental results away from the core of the draft tube ($0 \leq \bar{r} \leq 0.4$, $0.6 \leq \bar{r} \leq 1.0$). For the Part and High Load operating points, the magnitude of the experimentally measured velocity nearer the centre of the draft tube is significantly smaller than for the BEP. The flow is more still and, for the HL point, nearly reverses. At Part Load, the sliding mesh results under predict the velocity in the core. However, the thickening of the central region of lower velocity downstream of the cone shown from Line 1 to Line 2 is predicted correctly. At High Load, the velocity predicted with the sliding mesh simulations overestimates the experimental results, although, again the edges of the central region and the inflexion in the velocity profile at the edges of the central zones are predicted correctly. For the PL and HL conditions, overall the frozen rotor calculations give results which are closer to the experiments.

The calculated and measured radial velocity along lines 1 and 2 are compared in figure 10 below. The experimental data is discontinuous with peak and troughs, and is not symmetrical. The radial velocity magnitude is an order of magnitude lower than the axial velocity. The overall trends are correct in the numerical results, with the sliding mesh model giving slightly better results than the frozen rotor. However, for both models the calculated velocities show variations which are significantly larger than the variations in the measured velocities. A strongly refined mesh would probably be required in order to better resolve this very challenging region of the flow in the wake of the impeller blades, where velocities are low and variations of velocities of the same order of magnitude as the velocities themselves.

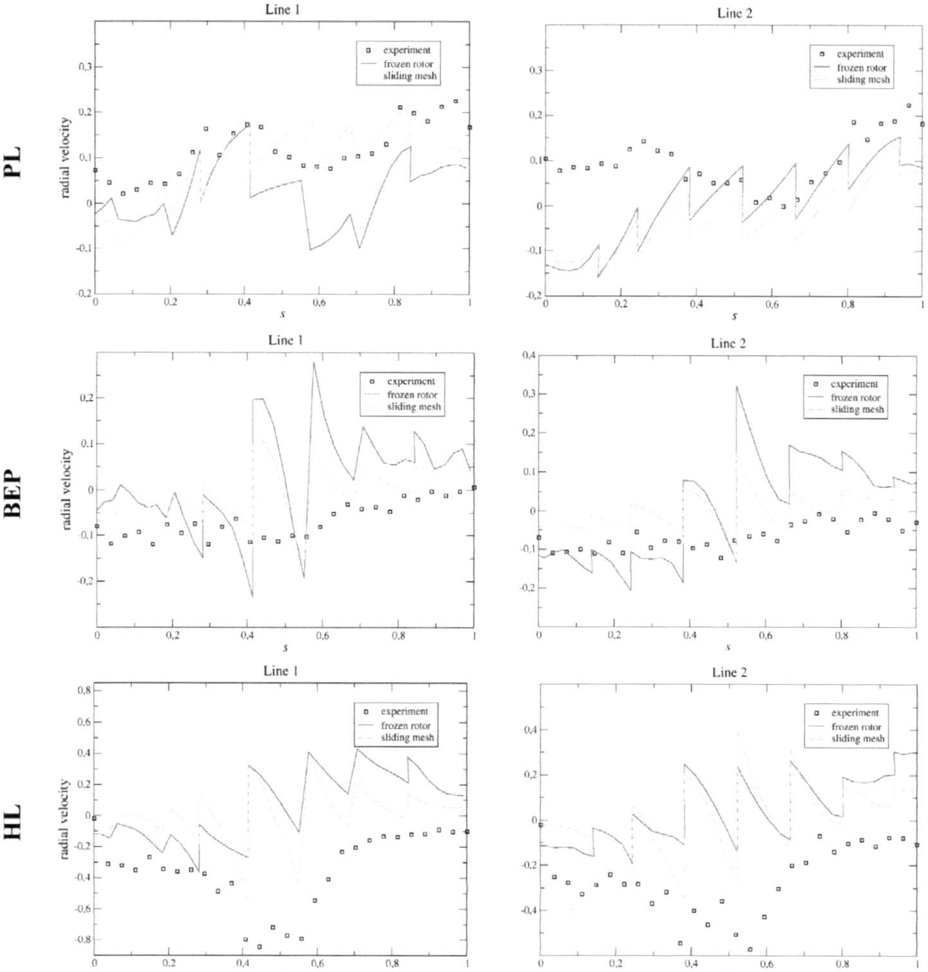

Figure 10: Numerical and experimental radial velocity vs. normalised radius along lines 1 and 2.

3.2. Pressure and Hydraulic Head

The net head is calculated as the total pressure change through the turbine, excluding the hydrostatic effects:

$$H = \frac{p_{abs,1} - p_{abs,2}}{\rho g} + \frac{v_1^2 - v_2^2}{2g} + \Delta z,$$

where Δz= 1.0715 corresponds to the height difference between the inlet and outlet reference planes (figure 11).

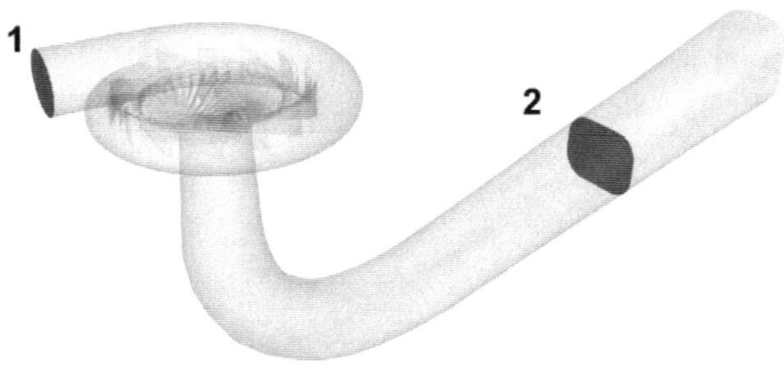

Figure 11: Reference planes used to compute net hydraulic head.

The calculated pressures and net hydraulic heads are compared to the experimental values for the three operating points in Tables 1-3 below.

Table 1: Pressure and hydraulic head. Part Load operation.

PL	VL2 [kPa] (error [%])	Pin [kPa] (error [%])	H [m] (error [%])
experiment	168.97	218.46	11.88
frozen rotor	161.60 (-4.36)	221.17 (+1.24)	12.18 (+2.53)
sliding mesh	161.51 (-4.41)	223.29 (+2.21)	12.35 (+3.96)

Table 2: Pressure and hydraulic head. Best Efficiency Point operation.

BEP	VL2 [kPa] (error [%])	Pin [kPa] (error [%])	H [m] (error [%])
experiment	173.17	215.93	11.97
frozen rotor	163.54 (-5.56)	217.20 (+0.59)	12.10 (+1.09)
sliding mesh	164.67 (-4.91)	219.49 (+1.65)	12.33 (+3.01)

Table 3: Pressure and hydraulic head. High Load operation.

HL	VL2 [kPa] (error [%])	Pin [kPa] (error [%])	H [m] (error [%])
experiment	178.87	212.94	11.88
frozen rotor	172.05 (-3.81)	212.51 (-0.20)	11.88 (-0.04)
sliding mesh	168.07 (-6.04)	216.58 (+1.71)	12.18 (+2.52)

The evolution with the different operating conditions and the overall magnitude of the pressure in the vaneless space upstream of the blades (VL2) are captured very well by both numerical models. Likewise, the inlet pressure and heads are very similar in the simulations and the experiments. Overall, the frozen-rotor results are in slightly better agreement with the experimental ones than the sliding mesh results.

Both models tend to overestimate the head, which can be explained by the fact that losses such as gap losses are neither modelled nor implicitly taken into account in the present CFD model. In fact, the larger values of hydraulic head computed with the sliding mesh model indicate that the numerical losses are smaller for the sliding mesh model than the frozen rotor one.

The pressure fluctuations in the draft tube (DT5 and DT6 sensors) were not captured in the computations, pointing to the inherent limitations of these models for such analyses.

4. Transient Operating Conditions

In addition to the simulations of the steady-state operation, work is also currently in progress to perform unsteady, frozen-rotor simulations of the transient operating conditions. The methodologies being developed and evaluated are based on the ALE solver already available in *Code_Saturne*, to impose displacements on the mesh nodes of and around the guide vanes.

In the first approach trialed, a global transformation was defined to prescribe the motion of all the nodes in a specific radial region, based on the desired variation of vane angle. Views of a proof of concept calculation are shown in figure 12 below for a simplified test case. Note that this global transformation does not coincide with the exact periodic transformation and leads to a deformation of the guide vanes.

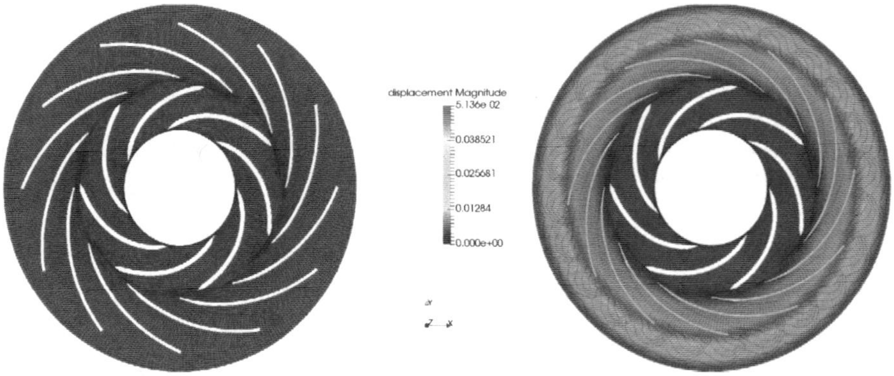

Figure 12: Validation of the mesh displacement for simplified stator. Displacement magnitude before (left image) and after (right image) displacement of the stator vane.

Nevertheless, the methodology was then applied to the Tokke case (load reduction, from BEP to PL) to evaluate if simulations could be carried with transient vanes motion. The results showed that the simulations were stable and results could be obtained, with the vanes moving over time, but that the nodes motion which had been prescribed resulted in excessive deformation of the vanes (figure 13).

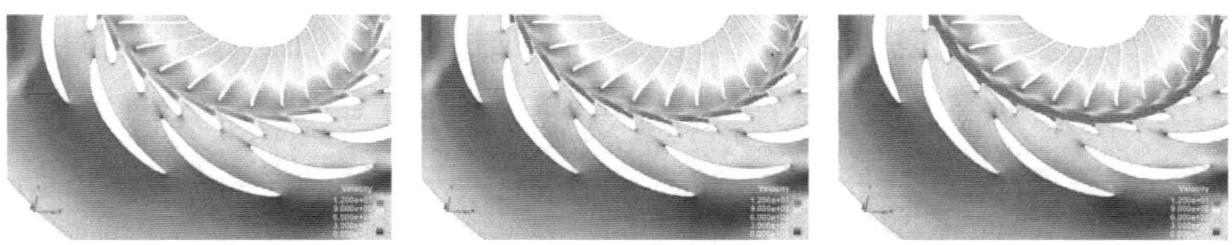

Figure 13: Snapshots of the transient vanes operation at three different vanes positions.

However, these results are encouraging and have provided the impetus for the development of a more robust approach based on new mesh-processing capabilities, to move individual nodes by interpolating their position over time between the initial and final mesh.

5. Conclusions

Modelling and simulations of the Tokke turbine at three steady-state operating points have been carried out using the open-source software SALOME for pre- and post-processing and *Code_Saturne* for the CFD simulations. The simulations on the full 3D model of the turbine have been run with two different methodologies, the frozen-rotor approximation and a moving mesh for the runner on over 200 compute cores.

Results obtained with the k-ω SST RANS model for turbulence show that the CFD solver can be used successfully to obtain an accurate local description of the flow at various operating conditions, without making changes to the computational model. The main characteristic of the turbine are also correctly estimated.

When considering which modelling approach to choose, the comparisons between the frozen rotor and sliding mesh results show that the sliding mesh approach does not offer significant advantages over the frozen rotor model for predictions of the flow in the draft tube and the global characteristic of the turbine. This is perhaps not overly surprising considering the large number of runner and guide vanes blades and the fact that stator/rotor interactions have a lesser impact in the draft tube. However, for more accurate simulations involving losses for example, or to look at the effect of rotor-stator interaction in detail for the different operating conditions, the sliding mesh modelling would be better suited. In the present state, the accurate prediction of pressure fluctuations appears to be beyond the scope of an incompressible flow solver.

The turbomachinery functionalities of *Code_Saturne* are already being used for hydraulic pump simulations. The reported simulations aimed at examining the capability of the code to simulate hydraulic turbine flows, including complex phenomena occurring away from design conditions and transient operations. The test results present an excellent opportunity to push and validate the solver and, whilst the results obtained are encouraging, they open the door to new developments for transient conditions and to performing more detailed flow analysis of such conditions and their impact on turbine design. Further work would also be of interest to better resolve the radial velocity component near the runner outlet.

Acknowledgements

The authors would like to thank the organisers of the Second Workshop for the use of their CAD, meshing, and experimental data.

References

[1] Goyal R, Bergan C, Cervantes MJ, Gandhi BK, Dahlhaug OG (2016c) Experimental investigation on a high head model Francis turbine during load rejection. 28th IAHR Symposium, Grenoble, France, July 4-8, 2016
[2] Bergan C, Goyal R, Cervantes MJ, Dahlhaug (2016) Experimental investigations of a high head model Francis turbine during steady-state operation at off-design conditions. 28th IAHR Symposium, Grenoble, France, July 4-8, 2016
[3] https://www.ntnu.edu/nvks/test-case
[4] SALOME, http://www.salome-platform.org/
[5] Archambeau F, Méchitoua N, Sakiz M. 2004 Code_Saturne: a Finite Volume Code for the Computation of Turbulent Incompressible Flows, *International Journal on Finite Volumes*; 1

On the relation between friction losses and pressure pulsations caused byRotor Stator interaction on the Francis-99 turbine

Petter T.K.Østby[1,2], Jan Tore Billdal[2], Bjørn Haugen[1], Ole Gunnar Dahlhaug[1]

[1]Norwegian University of Science and Technology, Trondheim, Norway
[2]Rainpower Norge AS, Norway

E-mail: petter.oestby@rainpower.no

Abstract. High head Francis runners are subject to pressure pulsations caused by rotor stator interaction. To ensure safe operation of such turbines, it is important to be able to predict these pulsations. For turbine manufacturers it is often a dilemma whether to perform very advanced and time consuming CFD calculations or to rely on simpler calculations to save development time. This paper tries to evaluate simplifications of the CFD model while still capturing the RSI phenomena and ensuring that the calculation does not underpredict the pressure amplitudes. The effects which turbulence modeling, wall friction, viscosity and mesh have on the pressure amplitudes will be investigated along with time savings with each simplification. The hypothesis is that rotor stator interaction is manly driven by inviscid flow and can therefore be modeled by the Euler equations.

1. Introduction

The ability to correctly predict the pressure pulsation due to rotor stator interaction (RSI) in high head Francis turbines has been the subject to research for a long time and the current state of the art methods have good correlation with experimental data [6]. However, when designing new turbines it is often desirable to quickly evaluate effects such as RSI without having to use advanced and time consuming methods. It is also desirable for the quick method to produce conservative results that do not underestimate the pressure pulsations. This paper evaluates the possibility to solve the flow equations without turbulence models and viscosity, in practice the Euler equations, while still modeling the underlying physics of RSI in a low specific speed Francis turbine. This is done by investigating the relation between friction losses in the guide vane domain and pressure pulsations observed in the runner caused by RSI.

The usage of the inviscid Euler equations to predict rotor stator interaction has been common in the aerospace field for decades. In in 1988, Giles [5] presented a numerical analysis of rotor-stator interaction in a low-speed gas turbine. For water turbines, the Euler equations have been used to predict stationary performance since the early 1990s[4, 6], but not for unsteady calculations. It is still common to use in-house Euler solvers to quickly evaluate design changes in a preliminary design phase.

1.1. Euler equations

The Euler equations (1) are a set of equations describing inviscid fluid flow and can be seen as a special case of the Navier-Stokes equations (2) with zero viscosity. Even though the Euler equations can be used for compressible calculations this paper refers only to the incompressible variant.

$$\frac{D\vec{v}}{Dt} = -\frac{\nabla p}{\rho} + \cancelto{0}{\mu\frac{\nabla^2\vec{v}}{\rho}} \qquad (1)$$

$$\frac{D\vec{v}}{Dt} = -\frac{\nabla p}{\rho} + \mu\frac{\nabla^2\vec{v}}{\rho} \qquad (2)$$

2. Analytical evaluation

The pressure pulsations due to RSI in the runner are, neglecting all acoustic effects, due to the spatially varying pressure field created by the guide vanes. As the guide vanes are obstructions in the flow field, their presence will cause variation in the pressure field. The magnitude of the amplitudes of the pressure field can, disregarding any turbulent and viscous effects, be assumed to be proportional to the velocity squared.

Ignoring the guide vanes for now one can evaluate the domain in which the guide vanes are located as a ring, figure 1, where the water enters at the greater radius and leaves at the smaller. In high head Francis turbines the velocity of the water entering the guide vane domain is often highly tangential. Ignoring friction, the flow can be described using the conservation of spin and conservation of mass which would imply a constant flow angle,α.

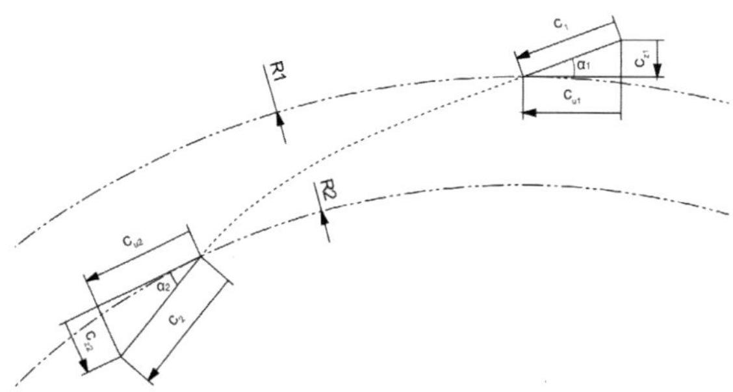

Figure 1. Velocity diagram through circular ring

$$c_u \cdot r = constant \tag{3}$$
$$c_z \cdot r = constant \tag{4}$$
$$\alpha = arctan(\frac{c_z}{c_u}) \tag{5}$$

However, including friction losses in this setup and applying conservation of torque(6) on the boundaries one can show that frictional losses in the tangential direction has to reduce the circumferential velocity, c_u.

$$\int_A (\vec{r} \times \vec{v} \cdot \rho) \cdot (\vec{v} \cdot \vec{n}) dA = \int_A (\vec{r} \times \vec{t}) dA \tag{6}$$

The inlet and outlet boundaries have a radial normal vector and the covers which cause friction have a normal vector in the axial direction. This, together with the assumption that the velocities are constant on the boundaries, equation (6) simplifies to (7), where M is the integral of tangential component of the friction force multiplied by its radius.

$$\dot{m} \cdot (r_1 \cdot c_{u1} - r_2 \cdot c_{u2}) = M \tag{7}$$
$$c_{u2} = c_{u1} \cdot \frac{r_1}{r_2} - \frac{M}{\dot{m} \cdot r_2} \tag{8}$$

Since the velocity of a high head Francis has a large tangential component in the guide vane domain, and thus a large M, the change in friction losses through the guide vane domain is expected to heavily influence the total velocity and thus the spatially varying pressure field. This suggests that a simplified calculation with less friction losses would be a conservative calculation with regards to pressure pulsations due to RSI.

The guide vanes can be included in such an analysis by assuming they behave like hydrofoils in a cascade with lift and drag. The drag will have the same effect as the friction in equation (6) reducing the circumferential velocity. Assuming that the guide vanes have a typical design then the lift will increase the circumferential velocity. Figure 2 shows a schematic representation of how the lift and drag affects the water. Since the lift coefficient is more or less independent of the viscosity and roughness while the drag coefficient is increased with viscosity and roughness [1], one can expect the presence of guide vanes to further enhance the effect fiction has on the circumferential velocity. An increase in friction with or without guide vanes will reduce the circumferential velocity and thus the spatially varying pressure field downstream.

IOP Conf. Series: Journal of Physics: Conf. Series **782** (2017) 012010 doi:10.1088/1742-6596/782/1/012010

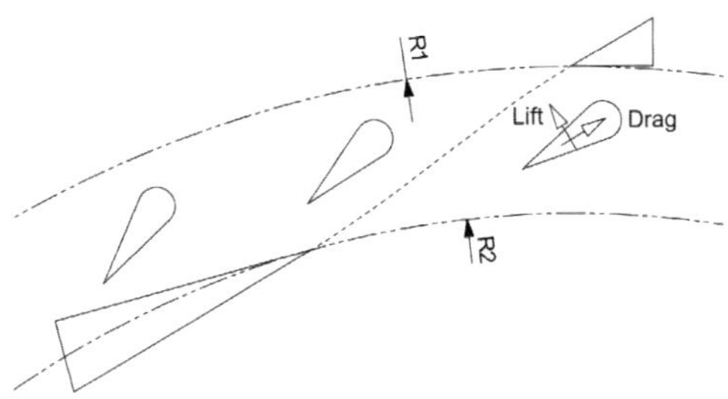

Figure 2. Velocity through circular ring with guide vanes

3. Reference case

The geometries used in this article are based on those provided with the Francis 99 workshop. The turbine is a low specific speed Francis turbine with 14 stay vanes, 28 guide vanes and a runner with 15 full vanes and 15 splitter vanes. The main parameters of the model turbine are given in table 1. Both the runner and the stationary components are equipped with pressure sensors which provide both steady and unsteady measurements. Only the on-board pressure sensors in the runner hub are used in this article. Their locations are shown in figure 3.

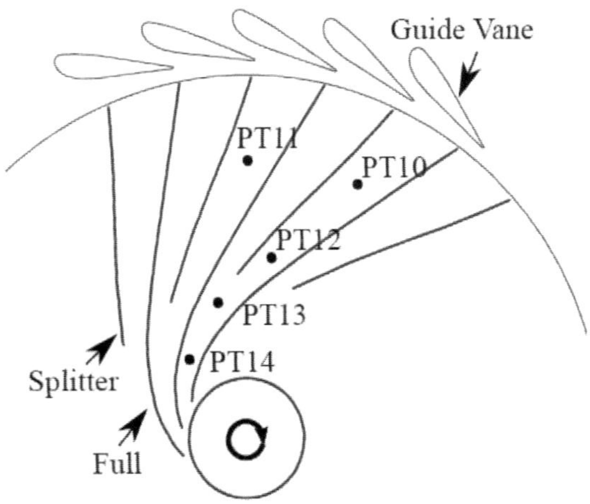

Figure 3. Runner Pressure Sensor Locations [2]

Table 1. Operating parameters of the model at best efficiency point [7]

Parameter	Symbol	Value	Unit
Head	H	12	m
Discharge	Q	0.2	m^3s^{-1}
Power	P	0.022	MW
Runner diameter	d	0.349	m
Runner speed	n	332	rpm
Reynolds number	Re	1.8×10^6	-

4. CFD evaluation

To verify the results from the analytical evaluation two tests were conducted with CFD. First a steady state simulation with only two guide vanes was run. Next a unsteady simulation was performed with the complete runner and all guide vanes. The simulations were performed with Ansys CFX v17.0 and the meshes were generated with Numeca's Autogrid. Different parameters were varied to investigate the relationship between friction losses and pressure pulsations. These changes also affect other parameters such as head, torque and efficiency of the turbine, but these effects are not reported as they are not the subject of this paper.

With the viscosity reduced to zero the system of equations effectively become the Euler equations. Ansys CFX can only solve the Navier-Stokes equations. There is no option to only solve the Euler equations. This means that even with a viscosity of zero the solver still solves the diffusion terms. For the viscous solutions the standard two equation k-ϵ turbulence model is used. The inviscid flow was modeled with turbulence modeling turned off.

4.1. Only guide vanes

4.1.1. Setup A set of simple CFD calculations were conducted including the guide vanes from the Francis-99 test. The setup included two guide vanesat the BEP opening. At the inlet a total pressure condition was used and a mass flow condition was used on the outlet. These conditions were chosen to avoid forcing a pressure field on either of the boundaries. The mass flow was chosen equal to the measured mass flow in the test rig at BEP. The mesh was an all hex mesh with 122880 elements and an average Y+30 with a dynamic viscocity of 10^{-3}.

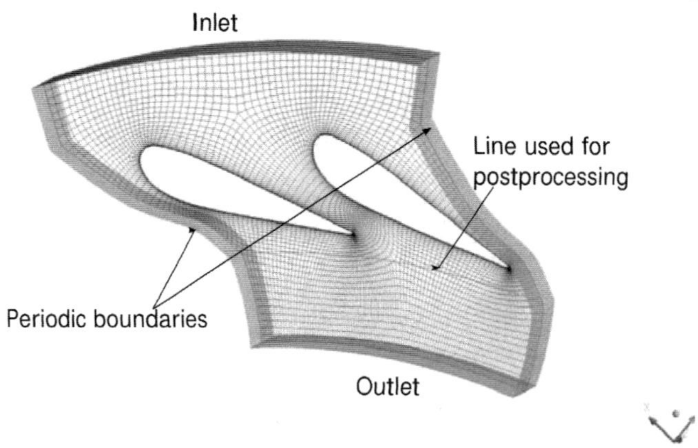

Figure 4. CFD setup for guide vane simulation

The base case was calculated with water, the $k-\epsilon$ turbulence model and hydraulicaly smooth walls with no-slip condition. Later this setup was modified to evaluate the relation between friction losses and pressure variations. All pressure variations are reported as the pressure amplitude at a line located at radius 319 mm, 10mm downstream of the guide vane trailing edge, which is shown in figure 4. Five different parameters were varied one at a time of the base case. The varied parameters are shown in table 2.

Table 2. Varied parameters for the guide vane tests

Parameter	Unit	Values
Main Models	[-]	k-ϵ,SST,laminar,slip/no-slip
Mesh	[# elements]	122880,288000,554240,858880
Dynamic Viscosity	$[\frac{kg}{m\,s}]$	$10^{-2}, 10^{-3}, 10^{-4}, 10^{-6}, 10^{-10}$
Wall Roughness	$[\mu m]$	1,6,12,24,48,100,200,400
Inlet flow angle	[]	15,20,25,30,35,40

4.1.2. Results The results show a clear relationship between the pressure amplitude and circumferential velocity, and also the pressure amplitude and the loss through the guide vane cascade. This is shown in figure 5.

The calculated pressure amplitudes depend mainly on parameters which also affects the circumferential velocity. This is in accordance with the findings of Antonsen [3], even though he reported less differences than what was found here. Changes in mesh and changes in inlet flow angle does not seem to affect neither the velocity nor the pressure amplitudes.

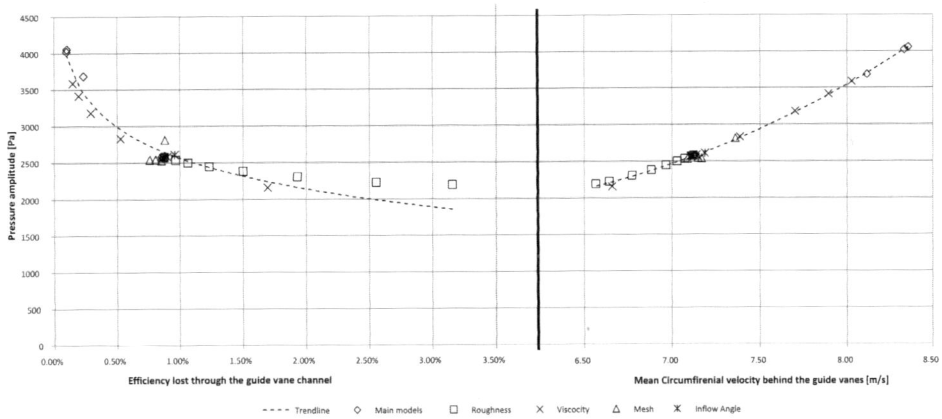

Figure 5. Spatially varying pressure amplitudes behind the guide vanes plotted against (right) local velocity and (left) efficiency lost

4.2. Runner and guide vanes

4.2.1. Setup The effect of the changes in amplitude for the spatially varying pressure field in the runner was evaluated using a full 360 unsteady calculation including both guide vanes and the runner. Three operating points were used in the calculations which are the same as the ones used in the experimental setup. The inlet boundary condition was set to a constant mass flow according to table 3, while the outlet had an average pressure constraint. Between the runner and guide vane domains a sliding rotor-stator interface was used. The total mesh contained 11.5 mio hex elements and the average Y+ values were 13 in the guide vane domain and 7 in the runner domain using regular water. 30 time steps were used per guide vane passing which equals 0.428 degrees per time step. CFX was required to achieve RMS-residuals of less than 10^{-5} before moving to the next time step. The number of iterations per time step was limited to maximum of 10.

As the variation in viscosity had the greatest impact on the results in the calculations with the guide vanes only, it was chosen as the main parameter to vary in the runner guide vane setup. The dynamic viscosity was varied with the following values $10^{-6}, 10^{-4}, 10^{-3}, 10^{-2}, 0[\frac{kg}{m\,s}]$. When the viscosity was set to zero, no turbulence model was used and slip conditions were applied to all walls. This is similar to using the Euler equations.

The same mesh is used for all values of viscosity. This was done to ensure that any changes in the pressure pulsations are the result of changes in viscosity. At the same time this may introduce an error as the boundary layer will be resolved differently due to Y+ changing with viscosity. The effect of the boundary layers are assumed to be minimal as the flow, especially near BEP, is without separations and recirculation and the automatic wall treatment of the CFX solver can handle large variations in y+.

Table 3. Operating points [2]

		PL	BEP	HL
Guide vane opening	[]	7.00	10.20	12.67
Mass Flow	[kg/s]	139.6	199.6	242.4
Rotational Speed	[rpm]	332.6	332.6	332.6

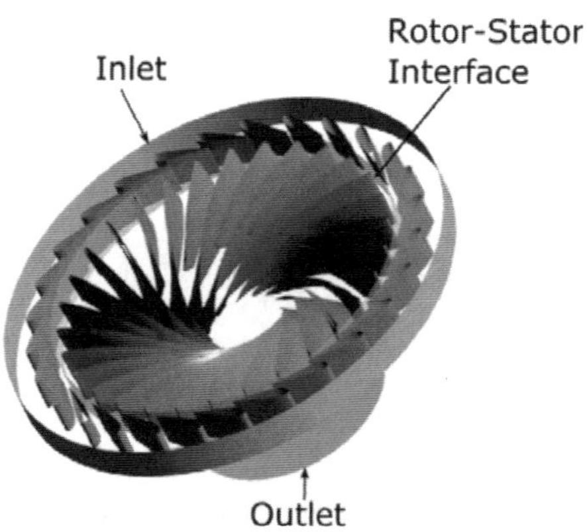

Figure 6. Runner and guide vane domain

4.2.2. Verification To ensure that a proper numerical setup was used, the calculated pressure amplitudes were compared to the measured pressure amplitudes provided by the pressure sensors in the runner hub. Both the phase shift of the time signal and amplitudes between the different sensors are compared to the CFD results. The time signal of the measured values is filtered through a band pass filter of 155 ± 5Hz. The mean and the 2.5 and 97.5 percentile values are used in the comparison. From the CFD a Fourier series at the guide vane passing frequency is run over the time signal to get the pressure amplitude and phase angle. CFD values are shown with uncertainty bands made from the 2.5 and 97.5 percentile of the values from all runner channels. Figure 8 shows that the phase angles corresponds with the measured values at the inlet but deviates a bit near the outlet. The pressure amplitudes at guide vane passing frequency are shown in figure 7. The pressure amplitudes are over estimated, on average, by 15% in the viscous CFD for all sensor locations. The best agreement between the viscous calculation and measurements is achieved at BEP (11%) while the results at PL (17%) and HL (21%) deviate more.

The calculation is very sensitive to guide vane opening as changes in guide vane opening at a given turbine discharge directly affects the circumferential velocity which again affects the pressure field.

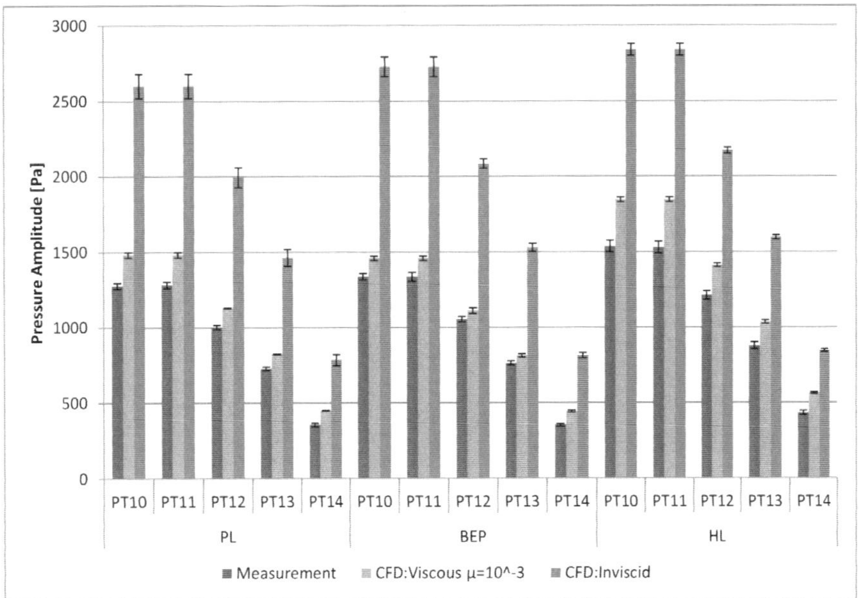

Figure 7. Comparison of measured and calculated pressure amplitudes

4.2.3. Results For each combination of operating point and dynamic viscosity one CFD calculation was run. The calculation runs until a steady and periodic solution is achieved. Then two more guide vane passings are simulated and a Fourier series is used on the pressure field in both the runner and guide vane domains. This extracts the amplitude and phase angle of the pressure field at the first harmonics of the guide vane passing frequency.

The instantaneous pressure field between the runner and guide vanes is shown at the High Load operating point in figure 10 plotted against the tangential position. It clearly shows the increase of the pressure variations with the inviscid solution while the general shape of the pressure field is more or less the same. It also clearly shows that the nodal diameter of the pressure field is ND2, which is visible as the pressure variation with two periods around the circumference. Both the blade passing frequency and the nodal diameter is predicted with the viscous and inviscid method.

How the pressure pulsations in the runner change with viscosity is shown in figure 9. Here the results are reported as the root mean square of the pressure amplitudes in the runner domain at the guide vane

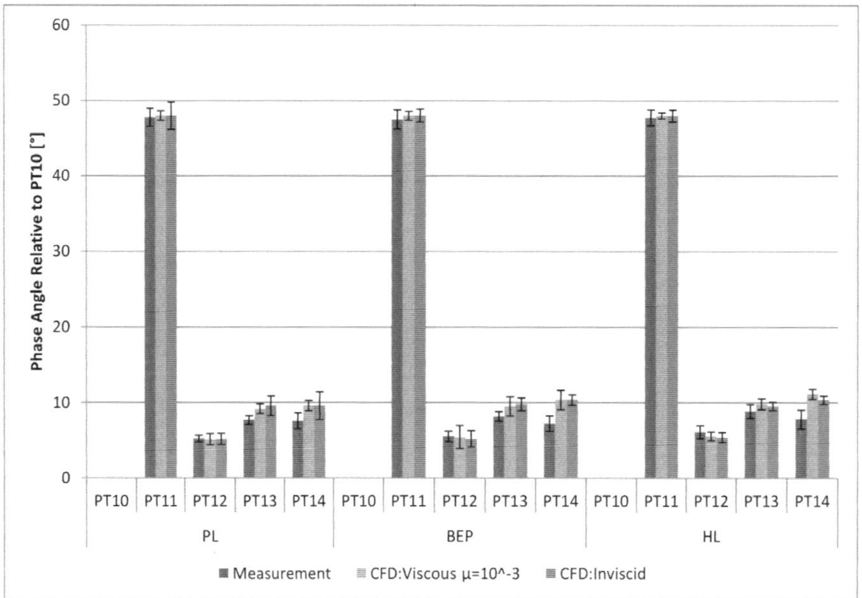

Figure 8. Comparison of measured and calculated phase angles

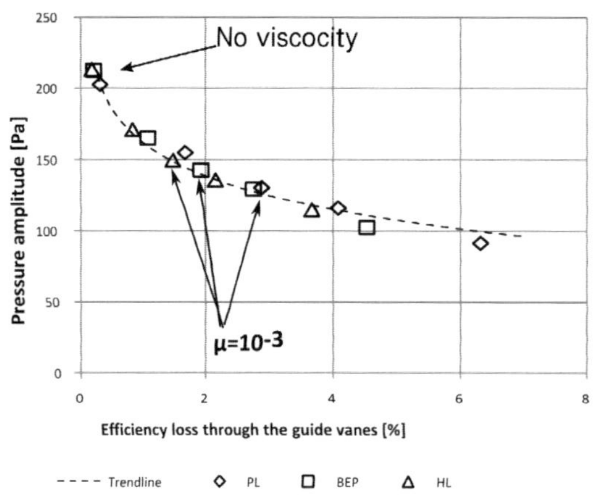

Figure 9. Average pressure amplitude in the runner vs guide vane losses

passing frequency plotted against the efficiency loss through the guide vane domain.

There are two interesting results shown in figure 9. First it shows that the pressure amplitudes can indeed be conservatively recreated using inviscous, laminar simulations. All calculated pressure amplitudes are significantly greater using the inviscid solution compared to the viscous solution. Also the trend of pressure amplitudes vs guide vane losses is very similar to the spatially varying pressure amplitudes vs guide

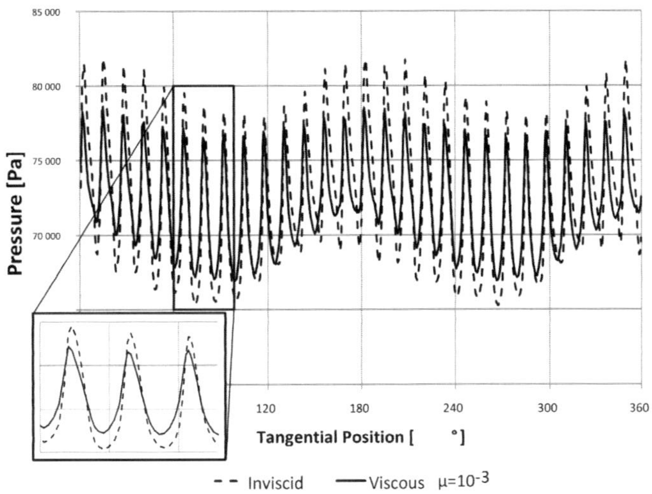

Figure 10. Instantaneous pressure field between runner and guide vanes at HL

vane loss shown in figure 5. The second interesting result from figure 9 is that changes in guide vane opening follows the same trendline as changes in viscosity . However, the increase in pressure amplitudes at greater guidevane openings is likely due to both the reduced spacing between the runner and guide vane and not only the change in friction losses and not the change in losses alone.

Another observation to be made from this is the relation between model and prototype pressure pulsations. For hydro turbines the Reynolds number is defined as in (9). When scaling a model to prototype the Reynolds number will increase. Since an increase in viscosity also is a reduction in Reynolds number, such a scaling will move the turbine left in figure 9. This means that the pressure amplitudes directly related to RSI without acoustic effects can be expected to increase slightly more than a direct scaling with head from a model test to prototype.

$$Re_u = \frac{d \cdot d \cdot \omega}{\nu} \tag{9}$$

5. Time saving

As the Euler equations have no diffusion terms and no turbulence equations, the time spent on each iteration is expected to be less than for using the Navier-Stokes equations. This is of course dependent on both the implementation of the equations in the solver and the problem at hand. Since CFX does not have a particular Euler solver, the equation solved is still the Navier-Stokes equation only with viscosity set to zero and without turbulence terms. Since slip-conditions are applied to the walls in the Euler case, the Y-plus requirements of the turbulence model is not present. However this has not been evaluated here as the mesh used is the same for both the inviscid and viscous solutions.

The calculations show that the average time spent calculating the unsteady 360° case was roughly 35% less using the laminar inviscid method compared to the turbulent viscous method. Examining on which equations the solver spends its time, it is clear that the time saved is a direct result of not solving the turbulence equations.

While a 35% reduction in time spent on each case is significant, having a pure Euler solver with a mesh tailored for the inviscid solver would most likely reduce the time even further.

6. Conclusions

Quickly producing conservative estimates for the pressure pulsations occurring inside of a low specific speed Francis turbine can be useful during a design phase. This paper has shown, both analytically and numerically, that there is a clear relation between friction losses in the guide vane cascade and the dynamic pressure variation observed in low specific speed Francis turbine calculated at the guide vane passing frequency. The observed pressure pulsations in the runner are inversly related to the friction losses.

Using conservation of momentum on a control volume around all guide vanes shows that increased friction losses have to reduce the tangential velocities which are propotional to the spatially varying pressure field squared. Observing the spatially varying pressure field behind the guide vanes from the rotating domain of the runner gives a transient pressure variation at the guide vane passing frequency. Together this shows that underestimating the friction losses, as with the inviscid Euler equations, will result in overestimation of the dynamic pressure field in the runner due to RSI.

Another observation which can be drawn from this is that pressure pulsations due to RSI increases with Reynolds number. This means that RSI-pressure pulsations observed at a modeltest will, when scaled to a prototype, increase more than a direct scaling by head.

Using the CFX solver with negligeble viscosity produced a pressure field both in the runner and in the vane less space which had similar waveforms and phase angles as the viscid solution. The pressure amplitudes were however between 50% and 100% greater compared to the fully viscid calculations. The time spent solving the inviscid flow field with out a turbulence model was found to be about 35% than the solving the viscous and turbulent flow field.

7. References

[1] Frank T. Abbott. *The effects of roughness at high Reynolds numbers on the lift and drag characteristics of three thick airfoils*. Langley Memorial Aeronautical Laboratory, 1944.

[2] Einar Agnalt. Pressure measurement inside a francis turbine runner. Master's thesis, NTNU, 2016.

[3] Øyvind Antonsen. *Unsteady flow in wicket gate and runner with focus on static and dynamic load on runner*. PhD thesis, NTNU, 2007.

[4] J. T. Billdal, Ø. Jacobsen, K. Bratsberg, H. I. Andersson, and H. Brekke. *Numerical Inviscid Flow Analysis of the GAMM Francis Runner*, pages 71–76. Vieweg+Teubner Verlag, Wiesbaden, 1993.

[5] MICHAEL B. GILES. Calculation of unsteady wake/rotor interaction. *Journal of Propulsion and Power*, 4(4):356–362, July 1988.

[6] Helmut Keck and Mirjam Sick. Thirty years of numerical flow simulation in hydraulic turbomachines. *Acta Mechanica*, 201(1):211–229, 2008.

[7] Chirag Trivedi, Michel J. Cervantes, and Ole G. Dahlhaug. Experimental and numerical studies of a high-head francis turbine: A review of the francis-99 test case. *Energies*, 9(2):74, 2016.